高等院校双语课精品教材

Materials in Civil and Construction Engineering

土木工程材料

Editor-in-Chief ◎ Yongjie Ding

Editors ◎ Xiaomei Bu, Xiaoyu Yang, Danni Li, Peng Guo

主　编 ◎ 丁勇杰

副主编 ◎ 卜晓梅　杨晓宇　李丹妮　郭　鹏

西南交通大学出版社

·成　都·

图书在版编目（CIP）数据

土木工程材料 = Materials in Civil and Construction Engineering: 英文 / 丁勇杰主编. —成都：西南交通大学出版社，2021.5
ISBN 978-7-5643-7943-8

Ⅰ. ①土… Ⅱ. ①丁… Ⅲ. ①土木工程 – 工程材料 – 高等学校 – 教材 – 英文 Ⅳ. ①TU5

中国版本图书馆 CIP 数据核字（2020）第 271462 号

土木工程材料
Materials in Civil and Construction Engineering

主编 丁勇杰

责任编辑 韩洪黎
封面设计 墨创文化

出版发行 西南交通大学出版社
（四川省成都市金牛区二环路北一段 111 号
西南交通大学创新大厦 21 楼）
邮政编码 610031
发行部电话 028-87600564 028-87600533
网址 http://www.xnjdcbs.com
印刷 四川森林印务有限责任公司

成品尺寸 185 mm × 260 mm
印张 9.25
字数 262 千
版次 2021 年 5 月第 1 版
印次 2021 年 5 月第 1 次
书号 ISBN 978-7-5643-7943-8
定价 35.00 元

课件咨询电话：028-81435775
图书如有印装质量问题 本社负责退换

PREFACE

A basic function of civil and construction engineering is to provide and maintain the infrastructure needs of society. Although some civil and construction engineers are involved in the planning process, most are concerned with the design, construction, and maintenance of facilities. The common denominator among these responsibilities is the need to understand the behavior and performance of materials. A basic understanding of the material selection process, and the behavior of materials, is a fundamental requirement for all civil and construction engineers performing design, construction, and maintenance.

First, this book introduces the basic properties of civil engineering materials. The introduction to the basic properties includes basic physical properties, mechanical properties and durability of materials. In addition, one of the responsibilities of civil and construction engineers is the inspection and quality control of materials in the construction process. This requires an understanding of material variability and testing procedures. The atomic structure of materials is covered in order to provide basic understanding of material behavior and to relate the molecular structure to the engineering response. So the second section presents the characteristics of the primary material types used in civil and construction engineering, such as building steel, aggregate, cement, masonry, concrete, wood, asphalt, and synthetic polymers. Since the discussion of concrete and asphalt materials requires a basic knowledge of aggregates, there is a chapter on aggregates. Moreover, since composites are gaining wide acceptance among engineers and are replacing many of the conventional materials, there is a chapter introducing composites.

Efforts have been made to make the teaching materials more

applicable, more substantial, more succinct, and more novel. Multiple sample problems have been added to each chapter to allow professors to vary assignments between semesters. Answering these questions and problems will lead to a better understanding of the subject matter.

As one of the civil engineering textbooks, this book is suitable for students majoring in "Civil Engineering", "Transportation Engineering", "Engineering Supervision", "Costing Engineering", and "Water Supply and Drainage Engineering", also for the engineers and technicians engaging in the relevant specialties. Due to the limited knowledge in the compilation of this book, mistakes and errors cannot be fully avoided. The comments and criticism from the readers will be highly appreciated.

Special thanks are due to the Southwest Jiaotong University Press project team, Ms. Xue Zhang, the chief editor, Ms. Wenyue Zhang, the editor, for their patience, understanding, and encouragement in publishing this manuscript.

The support from the Graduate Studies Office in the School of Civil Engineering and the Graduate School, at the Southwest Jiaotong University, is gratefully acknowledged.

Thanks to Jiangjuan Hao and Hao Yin for their contributions to the compilation of this book.

CONTENTS

Chapter 1 Basic Properties of Civil Engineering Materials

Chapter 2 Building Steel

Chapter 3 Aggregate

Chapter 4 Cement

Chapter 5 Portland Cement Concrete

Chapter 6 Masonry

Chapter 7 Asphalt and Asphalt Concrete

Chapter 8 Wood

Chapter 9 Synthetic Polymers

Chapter 1

Basic Properties of Civil Engineering Materials

The building and construction of civil engineering are composed of all kinds of civil engineering materials, which bear different load in different part. Therefore, civil engineering materials should have its corresponding basic properties. For example, good mechanical properties are needed for structural materials; wall materials should have the heat and sound insulation characteristics; roofing materials are waterproof and impermeable; pavement materials require anti-skidding and anti-abrasion properties. Moreover, a lot of civil engineering materials are exposed to atmosphere, and subjected to weathering by wind, rain, waves, ice and solar radiation, which means the durability is necessary.

The basic properties of civil engineering materials include physical property, mechanical property, durability, water resistance, fire-proof properties, decorativeness and so on.

1.1 Basic Physical Properties of Materials

1.1.1 Density, Apparent Density and Bulk Density

1. Density

Density is the mass per unit volume when the material is in the absolute dense state. It can be shown that:

$$\rho = \frac{m}{V}$$

In this formula: ρ is the density (g/cm^3);

m is the mass under dry conditions (g);

V is the volume under absolutely compact conditions (cm^3).

Volume in the absolute dense refers to the volume of the individual particles only (no voids). A lot of civil engineering materials include voids, such as brick, stone and concrete. As for these materials, it is required to grind the materials into powder, then dry to constant weight and measure its volume using Li Bottle, which is also called density bottle.

2. Apparent Density

Apparent density is the mass per unit volume when the material is in natural state.

$$\rho_0 = \frac{m}{V_0}$$

In this formula: ρ_0 is the apparent density (kg/m^3);

m is the mass under dry conditions (kg);

V_0 is the volume under natural conditions (m^3).

The volume in natural state includes the volume of the solid and internal pores. The apparent density varies with moisture content. Apparent density generally refers to apparent density in dry state.

3. Bulk Density

Bulk density is mass per unit volume when powdery or particle materials are in the stacking condition.

$$\rho_0' = \frac{m}{V_0'}$$

In this formula: ρ_0' is the bulk density (kg/m^3);

m is the mass under dry conditions (kg);

V_0' is the volume under packing conditions (m^3).

The volume in stacking state includes particle volume, inter-particle void volume, and internal pore volume. Bulk density is not an intrinsic property of a material; it varies from how the material is handled.

Table 1.1 Density, apparent density and bulk density of some civil engineering materials

Name	Density/(g/cm^3)	Apparent Density/(kg/m^3)	Bulk Density/(kg/m^3)
Steel	7.85	7850	-
Granite	2.6-2.9	2500-2850	-
Limestone	2.6-2.8	2000-2600	-
Gravels or Pebbles	2.6-2.9	-	1400-1700
Ordinary Sand	2.6-2.8	-	1450-1700
Sintered Clay Brick	2.5-2.7	1500-1800	-
Cement	3.0-3.2	-	1300-1700
Wood	1.55-1.60	400-800	-
Asphalt Concrete	-	2300-2400	-
Ordinary Concrete	-	2100-2600	-

1.1.2 Solidity Porosity and Voidage

Porosity is a fraction of the volume of voids over the total volume.

$$P=\left(1-\frac{V}{V_0}\right)\times 100\%=\left(1-\frac{\rho_0}{\rho}\right)\times 100\%$$

Porosity represents the densification of material. The higher porosity, the lower densification is. Porosity includes connected pore and closed pore according to its structure, and it can be classified into coarse pore, fine pore and micro pore according to its size.

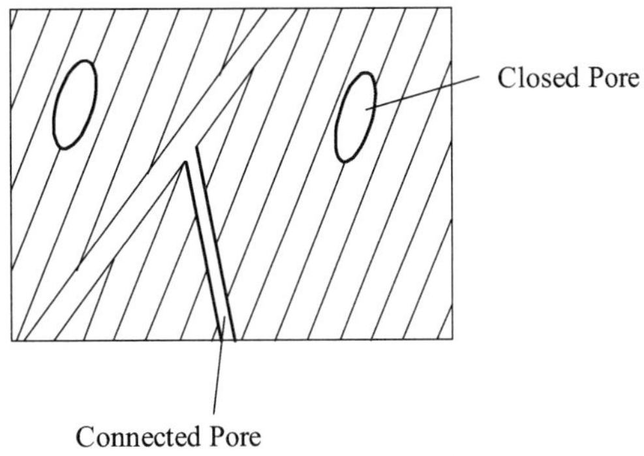

Figure 1.1 Sketch map of pores

Voidage is the proportion of spacing volume among the particles to the bulk volume of the non-coherent material in some container.

$$P'=\left(1-\frac{V}{V_0'}\right)\times 100\%=\left(1-\frac{\rho_0'}{\rho_0'}\right)\times 100\%$$

Voidage is an important parameter when controlling the gradation of concrete and calculating sand content.

1.1.3 Hydrophilic and Hydrophobic

Hydrophilic property refers to the material property which can be wetted when it contacts with water in the air ($0°\leqslant\theta\leqslant 90°$).

Most civil engineering materials belong to hydrophilic materials, such as stone, brick, block, glass and pottery. As for hydrophilic material, water-proof processing method can be used to improve its water resistance.

Hydrophobic property refers to the material property which cannot be wetted when it contacts with water in the air ($90°\leqslant\theta\leqslant 180°$). Asphalt, paraffin wax and some plastic used in civil engineering are hydrophobic materials.

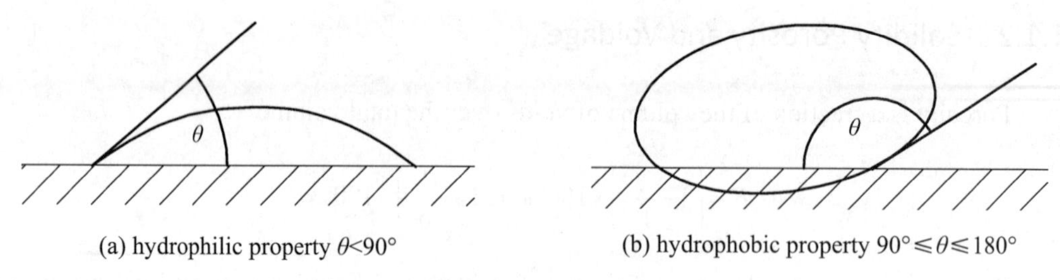

(a) hydrophilic property θ<90° (b) hydrophobic property 90°≤θ≤180°

Figure 1.2 Sketch map of wetting angle (θ)

1.1.4 Water Absorption and Moisture Absorption

Water absorption refers to the ratio of the weight of water absorbed by a material, to the weight of the dry materials.

Specific absorption of quality:

$$W_m = \frac{m_1 - m_2}{m_2} \times 100\%$$

Specific absorption of volume:

$$W_V = \frac{m_1 - m_2}{V} \times 100\%$$

In this formula: m_1 is the mass of the material at water-saturated state(g);

m_2 is the mass of the material under dry condition(g);

V is the volume of material under natural condition(cm^3).

The water absorption is related to porosity. Water can't enter into the dead-end pores. As for small interconnected pores, the more the pores are, the higher the water absorption is. Open pores are big, but it's difficult to store water, so its water absorption is less. Different material has diverse water absorption for its different internal structure.

Moisture absorption refers to the ratio of weight of materials with absorbed water in the moist air to that of dry materials.

$$W_m = \frac{m_1 - m_2}{m_2} \times 100\%$$

In this formula: m_1 is the quality of material in the moisture state(g);

m_2 is the quality of material under the dry condition(g).

1.1.5 Water Resistant and Waterproofing

Water resistant describes objects relatively unaffected by water or resisting the ingress of water under specified conditions. Such items may be used in wet environments or under water to specified depths. Waterproofing describes making an object waterproof or water-resistant (such as a camera, watch or mobile phone). "Water resistant" and "waterproof" often refer to penetration of water in its liquid state and possibly under pressure where damp proof refers to the resistance to humidity or dampness. In building construction, waterproofing is a fundamental aspect of creating a building envelope which is a controlled environment. The roof covering materials, siding, foundations, and all of the various penetrations through these surfaces need to be water-resistant and sometimes waterproof.

1.1.6 Anti-permeability

Anti-permeability refers to the property of something that cannot be pervaded by a liquid under pressure. Generally, Permeability coefficient or impermeability grade is used to describe the property. Permeability coefficient derives from Darcy's law.

Darcy's law at constant elevation is a simple proportional relationship between the instantaneous discharge rate through a porous medium, the viscosity of the fluid and the pressure drop over a given distance.

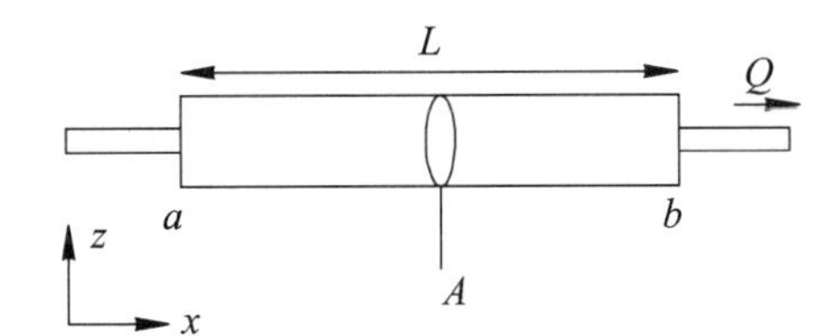

Figure1.3 Definitions and directions for Darcy's law

$$Q=\frac{-kA}{\mu}\frac{(p_b-p_a)}{L}$$

In this formula: Q is the total discharge (m^3/s);

k is the intrinstic permeability of medium;

A is the cross-sectional area of flow (m^2);

(p_b-p_a) is the total pressure drop (Pa);

μ is the viscosity (Pa · s);

L is the length over which the pressure drop is taking place (m).

The negative sign is needed because fluid flows from high pressure to low pressure.

As for concrete or mortar, impermeability grade is represented as the index of impermeability. Higher the grade is, better the impermeability is.

1.1.7 Freezing Resistance

Freezing resistance refers to the property of materials that can endure repeated freezing and thawing cycle without damage and its strength can't be obviously reduced. Generally, D_n is taken as frost resistance grade, in which n is the maximum times of freezing and thawing cycle when materials reach the regulated damage extent.

Both the anti-permeability and the freezing resistance are related to voidage of materials. Materials with less voidage or end voidage has higher anti-permeability and the freezing resistance. The small and connected pores are disadvantageous to these properties.

1.1.8 Thermal Conduction and Specific Heat

Thermal conduction is the transfer of heat from one part of a body to another or from one body to another through its physical contact. Nonmetals have a low coefficient of thermal conductivity. Metals have a much higher one because their free electrons transfer the vibrations much more rapidly. Thus, metals are good conductors of heat.

The rate of heat transfer by conduction is dependent on the temperature difference, the size of the area in contact, the thickness of the material, and the thermal properties of the material(s) in contact.

The quantity of heat transferred by conduction is defined as follows:

$$\lambda = \frac{Qd}{At(T_2 - T_1)}$$

In this formula: λ *is* the coefficient of thermal conductivity of the material [W/(m · K)];

Q is the conducted heat quantity (J);

A is the heat-transfer area (m^2);

t is the time for the heat transfer (s);

T_1 is the temperature on warmer side (K);

T_2 is the temperature on the colder side (K);

d is the thickness of a material (m).

The specific heat is the amount of heat per unit mass required to raise the temperature by one degree Celsius. The relationship between heat and temperature change is usually expressed in the form shown below where c is the specific heat. The relationship does not apply if a phase change is encountered, because the heat added or removed during a phase change does not change the temperature.

Table 1.2 lists a few construction materials and their thermal properties at nominal room temperature.

Table 1.2　Construction material thermal properties at room temperature[①-③]

Material	Thermal Conductivity /[W/(m・K)]	Specific Heat/[J/(kg・°C)]	Density/(kg/m^3)
Brick	0.7	840	1600
Concrete–cast Dense	1.4	840	2100
Concrete–cast Light	0.4	1000	1200
Granite	1.7-3.9	820	2600
Glass (window)	0.8	880	2700
Hardwoods (oak)	0.16	1250	720
Softwoods (pine)	0.12	1350	510
Polyvinyl Chloride	0.12-0.25	1250	1400
Paper	0.04	1300	930
Acoustic Tile	0.06	1340	290
Particle Board (low density)	0.08	1300	590
Particle Board (high density)	0.17	1300	1000
Fiberglass	0.04	700	150
Expanded Polystyrene	0.03	1200	50

1.2 Mechanical Properties of Materials

The mechanical properties of a material describe how it will react to external loads. Mechanical properties occur as a result of the physical properties inherent to each material, and are determined through a series of standardized mechanical tests.

1.2.1 Strength

Strength is the material capacity of resisting breakage by the external force. Strength includes compressive strength, tensile strength, shearing strength and bending strength according to the different form of external force.

The compressive strength is the capacity of a material or structure to withstand loads tending to reduce size. It can be measured by plotting applied force against deformation in a testing machine. Some material fracture at their compressive strength limit; others deform irreversibly, so a given amount of deformation may be considered as the limit for compressive

① Colorado Energy. R-Value Table (Insulation Values for Selected Materials) [OL]. [2019-07-16]http://www.coloradoenergy.org/procorner/stuff/r-values.html.
② Comfortable Low Energy Architecture[EB/OL]. [2002-11-01] http://www.new-learn.info/packages/clear/index.html.
③ JIM WILSON. "Thermal Properties Of Building Materials" [OL]. [2008-02-01] http://www.electronics-cooling.com/ 2008/02/thermal-properties-of-building-materials/.

load. Compressive strength is a key value for design of structures.

Tensile strength is the maximum stress that a material can withstand while being stretched or pulled before failing or breaking. Tensile strength is not the same as compressive strength and the values can be quite different. Some materials will break sharply, without plastic deformation, in what is called a brittle failure. Others, which are more ductile, including most metals, will experience some plastic deformation and possibly necking before fracture.

Shearing strength refers to a material's ability to resist forces that can cause the internal structure of the material to slide against itself. Shear strength is the maximum shear stress which a material can withstand without rupture. In structural and mechanical engineering the shear strength of a component is important for designing the dimensions and materials to be used for the manufacture/construction of the component (such as beams, plates, or bolts).[①]

Bending strength, or flexural strength is a material's ability to resist deformation under load. The flexural strength represents the highest stress experienced within the material at its moment of rupture. It is measured in terms of stress. Three and four points bend tests are commonly used to determine the flexural strength of a specimen.

Table 1.3 Formula of strength

Classification	Sketch map	Formula	Annotations
Compressive strength f_c	F F	$f_c = F/A$	f—Strength(MPa) F—Failure load(N) A—Loaded area(mm^2)
Tensile strength f_t	F F	$f_t = F/A$	
Shearing strength f_v	F F	$f_v = F/A$	
Bending strength f_m	l h b P/2 P/2 l	$f_m = 3Fl/2bh^2$ $f_m = Fl/bh^2$	

① Corrosion Pedia. Shear Strength [OL]. (2020-09-02) [2013-11-22] https://www.corrosionpedia.com/definition/1026/shear-strength.

1.2.2 Elastic and Plastic Deformation

In materials science, deformation is a change in the shape or size of an object due to an applied force or a change in temperature. A temporary shape change that is self-reversing after the force is removed, so that the object returns to its original shape, is called elastic deformation. Elastomers and shape memory metals such as nitinol exhibit large elastic deformation ranges, as does rubber. However, elasticity is nonlinear in these materials. Normal metals, ceramics and most crystals show linear elasticity and a smaller elastic range.

When a material distorts under pressure but does not return to its original shape after the pressure is released, it is called plastic deformation. This type of deformation is irreversible. However, an object in the plastic deformation range will first have undergone elastic deformation, which is reversible, so the object will return part way to its original shape. Soft thermoplastics have a rather large plastic deformation range as do ductile metals such as copper, silver, gold and steel, but cast iron does not. Hard thermosetting plastics, rubber, crystals, and ceramics have minimal plastic deformation ranges.

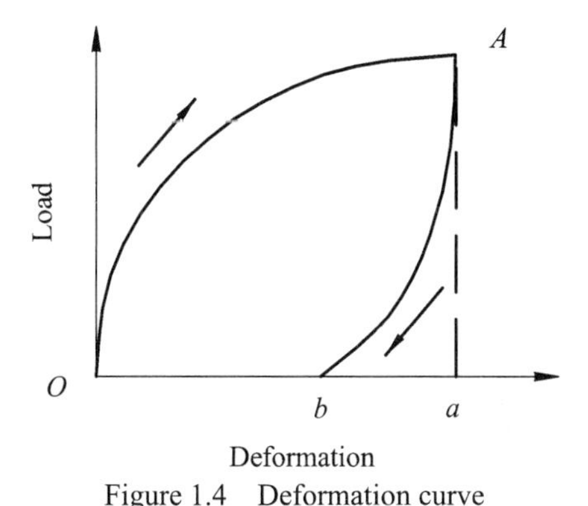

Figure 1.4 Deformation curve

1.2.3 Ductility and Brittleness

Ductility and brittleness are two of the most important physical properties of materials in construction engineering. Brittleness is the property of a material that will fracture without appreciable prior plastic deformation. Brittleness is lack of ductility and for a brittle material there is no plastic deformation. The elastic stage is followed by immediate fracture. Typical brittle materials include glass, concrete, ceramics, stone, and cast iron. Ductility is the property of a material that can be plastically deformed by elongation without fracture. Ductile materials can typically be plastically elongated with more than 15% before they fracture.

Typical ductile materials include copper, mild steel, and thermoplastics.[①]

Ductility of a material is its ability to deform when a tensile force is applied upon it. It is also referred to as the ability of a substance to withstand plastic deformation without undergoing rupture. Brittleness, on the other hand is exactly an opposite property of ductility as it is the ability of a material to break without first undergoing any kind of deformation upon application of force.[②]

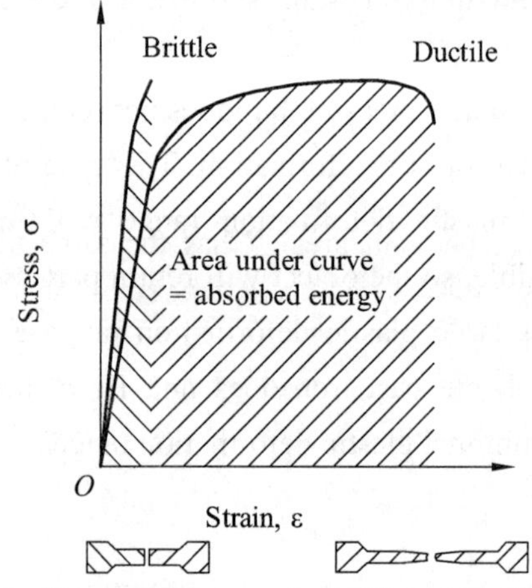

Figure 1.5 Stress-strain curves for brittle and ductile materials

1.2.4 Hardness

Hardness is a measure of how resistant solid matter is to various kinds of permanent shape change when a compressive force is applied. Hardness is dependent on ductility, elastic stiffness, plasticity, strain, strength, toughness, viscoelasticity, and viscosity.

Common examples of hard matter are ceramics, concrete, certain metals, and super-hard materials. There are three main types of hardness measurements: scratch, indentation, and rebound. Within each of these classes of measurement there are individual measurement scales. Scratch hardness tests are often used to determine the hardness of natural mineral. Steel, wood and concrete is usually determined by means of indentation hardness test. Rebound hammer measures the surface hardness of the concrete. The surface of concrete gets harder as concrete gains strength; thus, the strength of concrete can be estimated using this method.

---

① The Engineering Toolbox. Malleability, Brittlenes and Ductility [OL]. [2020-01-01] http://www.engineeringtoolbox.com/Malleability,-Brittlenes-and-Ductility-d_1851.html.

② OLIVIA. "Difference Between Ductility and Brittleness" [OL]. [2011-06-21] http://www.differencebetween.com/difference-between-ductility-and-vs-brittleness/.

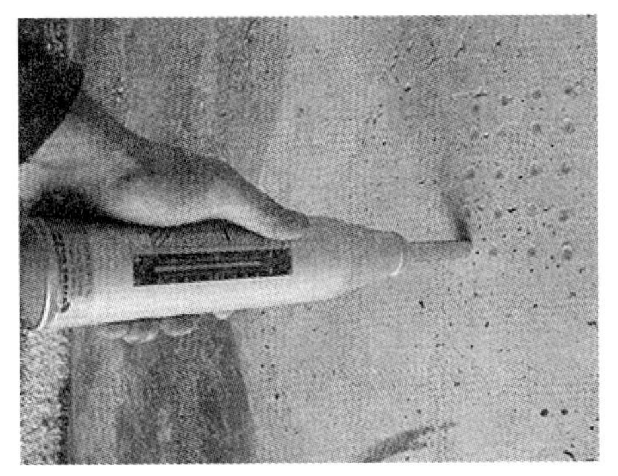

Figure 1.6 Concrete test hammer

1.2.5 Stress and Strain

Stress is defined as force per unit area. It has the same units as pressure, and in fact pressure is one special variety of stress. However, stress is a much more complex quantity than pressure because it varies both with direction and with the surface it acts on. Strain is defined as the amount of deformation an object experiences compared to its original size and shape.

It is unique for each material and is found by recording the amount of deformation (strain) at distinct intervals of tensile or compressive loading (stress). A lot of useful information about the material can be revealed by plotting the stress-strain diagram. Figure 1.7 shows the typical uniaxial tensile or compressive stress-strain curves for several engineering materials.

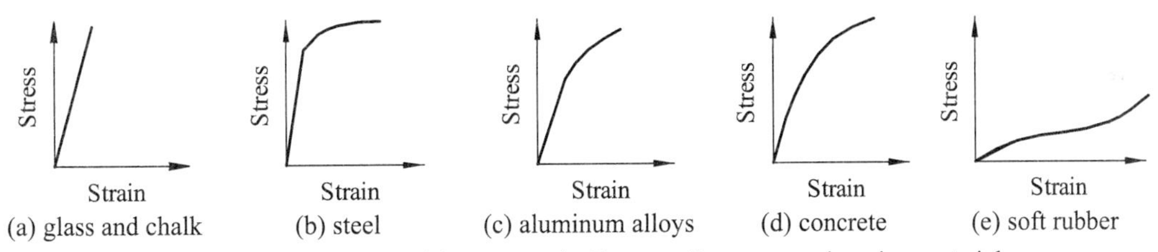

Figure 1.7 Typical uniaxial stress-strain diagrams for some engineering materials

1.3 Durability of Materials

Durability refers to ability to resist weathering action, chemical attack, abrasion, or any process of deterioration. A durable material helps the environment by conserving resources and reducing wastes and the environmental impacts of repair and replacement.

1.3.1 Durability of Concrete

Durability of concrete may be defined as the ability of concrete to resist weathering action, chemical attack, and abrasion while maintaining its desired engineering properties. Different concretes require different degrees of durability depending on the exposure environment and properties desired.

1.3.2 Mechanisms and Test Method

Table 1.4 shows important exposure conditions and deterioration mechanisms in concrete structures. In practice, several of these deterioration mechanisms can act simultaneously with possible synergistic effects.

Table 1.4 Exposure conditions and deterioration mechanisms in concrete structures

Durability Aspect/Exposure	Mechanism	Test Methods and other Standards
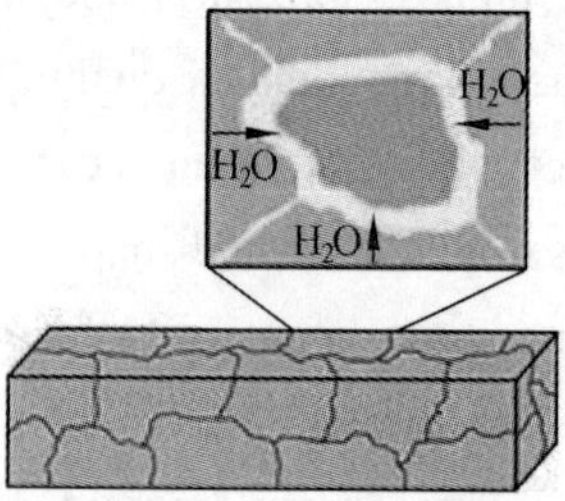 H_2O H_2O H_2O Alkali-Aggregate Reaction	Alkali-Silica Reaction Alkali-Carbonate Reaction	AASHTO PP65 ASTM C856 ASTM C1260 ASTM C1293 ASTM C1567
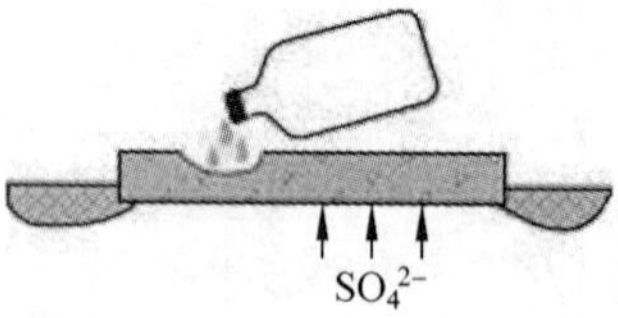 SO_4^{2-} Chemical Resistance	Sulfates DEF Seawater Acids	ASTM C1012 ASTM D516 ASTM C1582
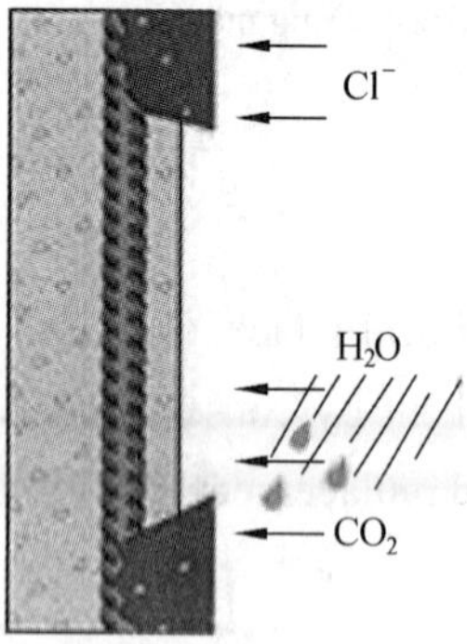 Cl^- H_2O CO_2 Corrosion of Reinforcement	Corrosion Corrosion Resistance Carbonation	ASTM C1202 AASHTO T 259 ASTM C1556 AASHTO T 260 ASTM C1152 ASTM C1218 ASTM C1524 AASHTO TP 11 AASHTO TP 22 AASHTO TP 26 AASHTO TP 55

Continued

Durability Aspect/Exposure	Mechanism	Test Methods and other Standards
Freeze-Thaw	Freezing and Thawing Deicer Scaling D-Cracking	ASTM C666 AASHTO T 161 AASHTO TP 18 ASTM C457Links ASTM C672
Miscellaneous	Abrasion Erosion Fire Resistance Efflorescence	ASTM C131 ASTM C535 ASTM C3744 ASTM C1137 AASHTO TP 58

Resistance to Alkali-Silica Reaction (ASR): ASR is an expansive reaction between reactive forms of silica in aggregates and potassium and sodium alkalis, mostly from cement, but also from aggregates, pozzolana, admixtures, and mixing water. The reactivity is potentially harmful only when it produces significant expansion.

Chemical Resistance: Concrete is resistant to most natural environments and many chemicals. Concrete is virtually the only material used for the construction of wastewater transportation and treatment facilities because of its ability to resist corrosion caused by the highly aggressive contaminants in the wastewater stream as well as the chemicals added to treat these waste products.

Resistance to Sulfate Attack: Excessive amounts of sulfates in soil or water can attack and destroy a concrete that is not properly designed. Sulfates (for example calcium sulfate, sodium sulfate, and magnesium sulfate) can attack concrete by reacting with hydrated compounds in the hardened cement paste. These reactions can induce sufficient pressure to cause disintegration of the concrete.

Chloride Resistance and Steel Corrosion: Chloride present in plain concrete that does not contain steel is generally not a durability concern. Concrete protects embedded steel from corrosion through its highly alkaline nature.

Resistance to Freezing and Thawing: The most potentially destructive weathering factor is freezing and thawing while the concrete is wet, particularly in the presence of deicing chemicals. Deterioration is caused by the freezing of water and subsequent expansion in the paste, the aggregate particles, or both.

Abrasion Resistance: Concrete is resistant to the abrasive effects of ordinary weather. Examples of severe abrasion and erosion are particles in rapidly moving water, floating ice, or areas where steel studs are allowed on tires. Abrasion resistance is directly related to the strength of the concrete.

Questions

1.1 What are density, apparent density and bulk density? Explain their differences.

1.2 What are porosity and voidage? How to calculate? Briefly describe the relationship between them.

1.3 What is water absorption? What are the influencing factors of water absorption?

1.4 What is strength of a material? How to calculate various kinds of strength according to the different form of external force?

1.5 Describe the differences between elastic deformation and plastic deformation.

1.6 What is durability of concrete? Describe the text method of durability.

References

[1] INCROPERA, FRANK P, DEWITT, DAVID P. Introduction to Heat Transfer[M]. 2nd ed. New York: John Wiley and Sons, 1990.

[2] MICHAEL S MAMLOUK, JOHN P ZANIEWSKI. Materials for Civil and Construction Engineers [M]. Stockton: QWE Press, 2005.

Chapter 2
Building steel

2.1 Introduction

Building steels are for engineering buildings, including profiled bars, armor plates, steels and steel wires. Building steel is the material produced under strict technical conditions, and it has the following advantages: high strength-weight ratio, better plasticity and toughness, flexible processing and the properties to bear impacts and vibration loads; the disadvantages are: easy to be corroded and high cost of repairs.

These characteristics determine that steel is one of the important materials needed by economic construction departments. In construction, the steel structures consisted by steel in various shapes have high security and light deadweight, used for large-span and high-rise structures. But though concrete structures have heavy deadweight, the usage of steel is decreased greatly, and it can overcome the corrosion and high cost of repairs of steel. Thus, steel is widely used in concrete structures.

This chapter focuses on the properties of building steel and it introduces the standards and selection of building steel. It simply introduces corrosion of building steel and the measures to prevent corrosion.

2.1.1 Steel

Steel is an alloy of iron with typically a few percent of carbon to improve its strength and fracture resistance compared to iron. Many other additional elements may be present or added. Stainless steels that are corrosion and oxidation resistant need typically an additional 11% chromium. Because of its high tensile strength and low cost, steel is used in buildings, infrastructure, tools, ships, trains, cars, machines, electrical.

The carbon in typical steel alloys may contribute up to 2.14% of its weight. Varying the amount of carbon and many other alloying elements, as well as controlling their chemical and physical makeup in the final steel (either as solute elements, or as precipitated phases), slows the movement of those dislocations that make pure iron ductile, and thus controls and enhances its qualities. These qualities include the hardness, quenching behavior, need for annealing, tempering behavior, yield strength, and tensile strength of the resulting steel. The increase in steel's strength compared to pure iron is possible only by reducing iron's ductility.

Steel was produced in blooming furnaces for thousands of years, but its large-scale, industrial use began only after more efficient production methods were devised in the 17th century, with the introduction of the blast furnace and production of crucible steel. This was followed by the open-hearth furnace and then the Bessemer process in England in the mid-19th century. With the invention of the Bessemer process, a new era of mass-produced steel began. Mild steel replaced wrought iron.

Further refinements in the process, such as basic oxygen steelmaking (BOS), largely replaced earlier methods by further lowering the cost of production and increasing the quality of the final product. Today, steel is one of the most common manmade materials in the world, with more than 1.6 billion tons produced annually. Modern steel is generally identified by various grades defined by assorted standards organizations.

2.1.2 Classification

1. By Smelting Processes

During smelting, the removal degrees of impurities by different smelting methods are not the same, so the steel qualities are different. Recently, there are three kinds of steel, including Bessemer steel (converter steel), Siemens-Martin steel, and electric steel.

(1) Bessemer Steel

The smelting process of this steel is to use the molten pig iron as the raw material without any fuel and to make steel with air being blown through the molten iron (the raw material) from the bottom or the sides of the converter, called pneumatic converter steel; if pure oxygen is used to replace the air, it is called the oxygen converter steel. The disadvantage of pneumatic converter steel is that the nitrogen, hydrogen and other impurities in the air will interfuse easily, the smelting time is short, and the impurity content is difficult to control, so the quality is poor; the quality of oxygen converter steel is high, but the cost is a little higher.

(2) Siemens-Martin Steel

The process of Siemens-Martin steel is to use solid or fluid pig iron, ore or waste steel as the raw materials and coal gas or heavy oil as the fuel and to remove the impurities from the iron by oxidation with the oxygen in ore or waste steel or the oxygen being blown through the iron. Because the smelting time is long (4-12 h), the impurities are removed clearly and the quality of steel is good. But the cost is higher than that of Bessemer steel.

(3) Electric Steel

The process of electric steel is to make steel by electric heating. The heat source is high-tension arc, and the smelting temperature is high and can be adjusted freely, so the impurities can be removed clearly and the steel quality is good.

2. By Press-working Modes

In the process of smelting and ingot-casting, there will be uneven structures, foams or other defects happening to the steel, so the steel used in industry should be processed by press to eliminate the defects. Meanwhile, there is requirement for shapes. The press-working modes include hot working and cold working.

(1) Hot-working steel: Hot working is to heat the steel ingot to a certain temperature and to conduct press-working to the steel ingot in the plastic state, such as hot rolling and hot forging.

(2) Cold-working steel: The steel is processed under the normal temperature.

3. By Chemical Elements

Steel Classifications (GB/T 13304-1991), the Chinese standard, recommends two classification methods: one is to classify by chemical elements, and the other is to classify by quality degrees. By chemical elements, there is non-alloy steel, lean-alloy steel and alloy steel.

(1) Non-alloy steel: that is carbon steel with few alloy elements.

(2) Lean-alloy steel: that is the steel with low alloy elements.

(3) Alloy steel: that is the steel added with more alloy elements to improve some properties of the steel.

4. By Quality Degrees

According to quality degrees, the steel can be classified into: common steel, quality steel and advanced quality steel.

5. By purposes

The steel can be classified by purposes, such as construction steel, railway steel, and pressure vessel steel. The construction steel can be classified by purposes into the steel for steel structures and that for concrete structures. At present, the steel commonly used in constructions includes carbon structural steel and lean-alloy and high-strength structural steel.

2.2 Properties of Building Steel

The essential properties of steels in steel structure and reinforced concrete in civil engineering are:

(1) Mechanical property: tensile strength, impact toughness, fatigue strength.

(2) Processing property: cold bending and welding property.

2.2.1 Mechanical Properties

1. Tensile Strength

It is the most important property of the building steels. The tensile strength of construction steel includes: yield strength, ultimate, tensile strength, and fatigue strength.

(1) Low Carbon Steel Stress-strain Curves (Figure 2.1)

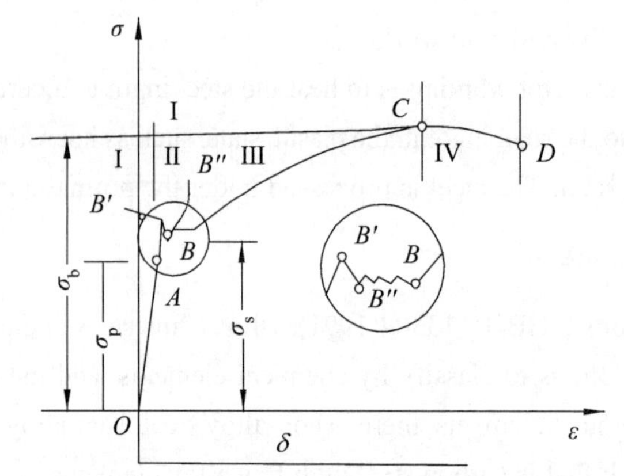

Ⅰ-The elastic stage, expressed by σ_p; Ⅱ-The yield stage, expressed by σ_s;
Ⅲ-The reinforcement stage, expressed by σ_b; Ⅳ-The necking stage.

Figure 2.1 Low carbon steel stress-strain curves

(2) Strength

① Yield Strength

As shown in Figure 2.1, at the yield stage, the corresponding stress of the highest point on the hackle is called the upper yield point (*B*); get rid of the initial transient effect, the corresponding stress of the lowest point is called the lower yield point (*B'*). The Chinese standard regulates that the stress of the lower yield point is the yield strength of the steel, expressed by σ_s.

When the actual stress of a structure reaches the yield point, there will be irretrievable deformation which is not allowed in constructions. Thus, yield strength is the main base to determine the allowable stress of the steel.

② Ultimate Tensile Strength (Simply Called Tensile Strength)

It is the ultimate tensile stress that the steel can bear under the role of tension, shown in Figure 2.1, the highest point of stage Ⅲ. σ_s/σ_b can reflect the availability and safety of steel. The smaller the yield ratio, the more reliable the structure is. However, if the ratio is too small, the available utilization ratio of the steel will be too low, and the reasonable yield ratio should lie between 0.6-0.75. Therefore, the yield strength and the tensile strength are the major test indexes of the mechanical properties of steel.

③ Fatigue Strength

Under the role of alternating loads, steel will be damaged suddenly when the stress is far below the yield strength, and this damage is called fatigue failure. The value of stress at which failure occurs is called fatigue strength, or fatigue limit. The fatigue strength is the highest value of the stress at which the failure never occurs. Generally, the biggest stress that the steel bears alternating loads for 106-107 times and no failure occurs is called the fatigue strength.

2. Plasticity

The construction steel should have good plasticity. In projects, the plasticity of the steel is usually expressed by the elongation. Elongation refers to the ratio of the increment of the gauge length to the original gauge length when the specimen is stretched off, expressed by δ(%), shown in Figure 2.2.

$$\delta = \frac{L_1 - L_0}{L_0} \times 100\%$$

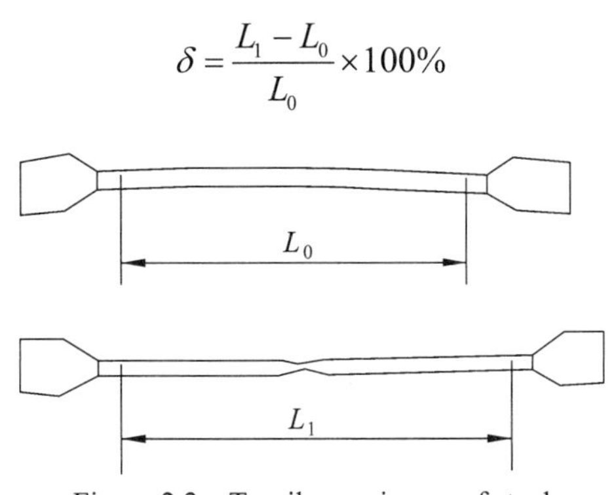

Figure 2.2 Tensile specimens of steel

Plasticity is an important technical property for steel. Though the structures are used during the elastic stage, the part where the stress converges could be beyond the yield strength. And certain plasticity can guarantee the redistribution of the stress to avoid failure of structures.

3. Impact Durability

Impact durability refers to the property that the steel resists loads without being damaged. It is regulated that the impact durability is expressed by the work spent on the unit area of the damaged notch when the standard notched specimen is stricken by the pendulum of the impact test, with the sign α_K, and the unit J, as shown in Figure 2.3. The bigger α_K is, the more work will be spent in damaging the specimen, or the more energy the steel will absorb before getting cracked, and the better the durability of the steel is.

The impact durability of the steel is related to its chemical elements, smelting, and processing. Generally, *P* and *S* contents in steel are high, and impurities and the tiny cracks forming in smelting will lower the impact durability.

In addition, the impact durability of the steel can be influenced by temperature and time. At the room temperature, the impact durability will decline little with the temperature falling, and the damaged steel structure reveals the ductile fracture; if the temperature falls into a range, α_k declines suddenly, the steel reveals the brittle fracture, and the temperature is very low when cold brittle fracture occurs. In north, especially the cold places, the brittle fracture of the steel should be tested when the steel is used. The critical temperature of its brittle fracture should be lower than the lowest temperature of the place. Because the measurement of the critical temperature is complicated, what is regulated in standards is the impact values at the negative temperature -20 °C or -40 °C.

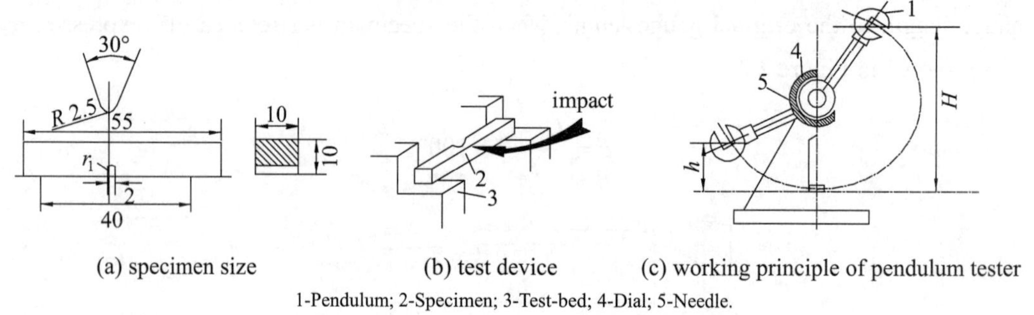

(a) specimen size (b) test device (c) working principle of pendulum tester

1-Pendulum; 2-Specimen; 3-Test-bed; 4-Dial; 5-Needle.

Figure 2.3 The test principle of impact durability

4. Rigidity

Rigidity is the property to resist the plastic deformation when there is a hard object press into the steel within the partial volume of the surface, often related to the tensile strength. Recently, there are various methods to measure the rigidity of the steel, and the most common one is Brinell hardness, expressed by HB.

2.2.2 Process Properties

1. Cold Bending Property

Cold bending is the property that the steel bears the bending deformation under the normal conditions. The cold bending is tested by checking whether there are cracks, layers, squamous drops and ruptures on the bending part after the specimen goes through the regulated bending.

Generally, it is expressed by the ratio of the bending angle α and the diameter of the bending heart d to the thickness of the steel or the diameter of the steel a. Figure 2.4 shows that the bigger the bending angle is, the smaller the ratio of d to a is, and the better the cold bending property is.

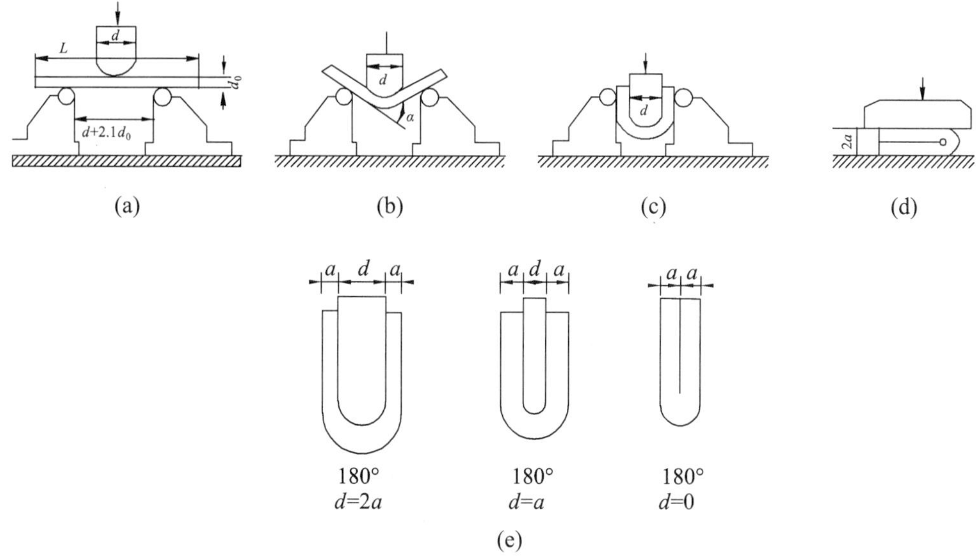

Figure 2.4 Cold bending test of steel

Cold bending test is helpful to expose some defects of steel, such as pores, impurities and cracks. In welding, the brittleness of parts and joints can be found by cold bending test, so the cold bending test is not only the index to check plasticity and machinability, but also an important index to evaluate the welding quality. The cold bending test for the steel used in important structures or the bended steel should be qualified.

2. Weld Ability

Welding is the major mode for the combination of steel. The quality of welding depends on the welding techniques, welding materials and the weld ability of steel.

Weld ability refers to the property that under a certain welding condition, there is no crack or hard rupture in or around welding seams and the mechanical property after welding, especially the strength, should be not lower than the original one.

Weld ability is often impacted by chemical components and the contents. The weld ability will decrease, if the carbon content is more than 0.3%, or there is more sulfur, or the impurity content is high, and the alloy elements content is high.

Usually, the steel used for welding is the oxygen converter or the Siemens-Martin fully-killed steel with lower carbon content. For the high carbon steel and alloy steel, preheating and heat treatment should be adopted respectively before and after welding in order to improve the hard brittleness of the steel after welding.

2.2.3 Affecting Factors of Properties

1. Influences of Chemical Elements

(1) Carbon

Carbon is the major element that determines the properties of steel. With the increasing of carbon content, the rigidity and the strength of steel will increase, and its plasticity and toughness will decrease. If the carbon content is more than 1 %, the ultimate strength of the steel begins to fall. In addition, if the carbon content is too high, the brittleness and aging sensitivity of the steel will rise, which reduce its ability to resist the corrosion of the atmosphere and weld ability.

(2) Phosphor and Sulfur

Phosphor is similar with carbon that can improve the yield point and bending strength of steel, lower its plasticity and toughness, and greatly increase its cold brittleness. But the segregation of phosphor is serious and there are cracks in welding, so phosphor is one of the elements that can lower the weld ability of steel. Thus, in carbon steel, the phosphor content should be controlled strictly; but in alloy steel, it can improve the resistance to atmospheric corrosion of steel, and can also be the alloy element.

In steel, sulfur exists in the mode of FeS. FeS is a kind of low melting compound that has been melted when the steel is processed or welded in the state of glowing red and will lead to cracks inside the steel, called hot brittleness. The hot brittleness greatly reduces the process ability and weld ability of steel. In addition, the segregation of sulfur is serious that can reduce the impact-resistance, fatigue strength and anti-corrosion of steel. Thus, the sulfur content should also be controlled strictly.

(3) Oxygen and Nitrogen

Oxygen and nitrogen can partly dissolve in ferrite and most of them exist in the mode of compounds. These non-metals contain impurities that reduce the mechanical properties of steel, especially the toughness of steel, and can accelerate aging and lower weld ability. Thus, the oxygen and nitrogen should be controlled strictly in steel.

(4) Silicon and Manganese

Silicon and manganese are the elements added purposely during steelmaking for deoxygenation and desulphurization. Because silicon can combine with oxygen greatly, it can capture the oxygen in ferric oxide to generate silicon dioxide and stay in the steel slag. Most of the remaining silicon will dissolve in ferrite. And when the content is low (less than 1%), it can improve the strength of steel and has little influence on plasticity and toughness.

Combining force of manganese with oxygen and sulfur is higher than that of iron, so manganese can change FeO and FeS into MnO and MnS respective 19 and stay in the steel slag. And the remaining manganese dissolves in ferrite and twists the crystal lattice to prevent slippage and deformation, greatly improving the strength of steel.

2. Influences of Cold Working and Heat Treatment

(1) Cold Working

Cold working is the process that steel is processed at the room temperature. The common cold working modes for construction steel include: cold stretching, cold drawing, cold rolling, cold twisting, notching.

At the room temperature, beyond the elastic range of the steel, the plastic deformation strength and rigidity of the steel have increased and its plasticity and toughness have decreased, which is called cold-working strengthening. It improves the strength and the yield strength can raise 20%-30% after cold working. Within a certain range, the bigger the cold-working deformation is, the greater the yield strength increases, and the more the plasticity and the toughness decrease.

(2) Heat Treatment

Steel can be hardened or softened by using heat treatment; the response of steel to heat treatment depends on its alloy composition. The basic process is to heat the steel to a specific temperature, hold the temperature for a specified period of time, then cool the material at a specified rate. Common heat treatments employed for steel includc anncaling, normalizing, hardening, and tempering.

The objectives of annealing are to refine the grain, soften the steel, remove internal stresses, remove gases, increase ductility and toughness, and change electrical and magnetic properties. Four types of annealing can be performed, depending on the desired results of the heat treatment: full annealing, process annealing, stress relief and spheroidization.

Normalizing is similar to annealing, with a slight difference in the temperature and the rate of cooling. It produces a uniform, fine-grained microstructure. However, since the rate of cooling is faster than that used for full annealing, shapes with varying thicknesses results in the normalized parts having less uniformity than could be achieved with annealing.

Steel is hardened by heating it to a temperature above the transformation range and holding it until austenite is formed. The steel is then quenched (cooled rapidly) by plunging it into, or spraying it with, water, brine, or oil. Due to the rapid cooling, hardening puts the steel in a state of strain. This strain sometimes causes steel pieces with sharp angles or grooves to crack immediately after hardening. Thus, hardening must be followed by tempering.

The predominance of martensite in quench-hardened steel results in an undesirable brittleness. Tempering is performed to improve ductility and toughness.

2.3 Standards and Selection

2.3.1 The Steel Used for Steel Structures

1. Carbon Structural Steel

It refers to the structural steels of various steel structure, hot-rolled sheet and section steels used in engineering.

(1) Grade and Representation

Carbon Structural Steel (GB 700-88), the national standard, regulates that grade consists of the letter of yield point, the value of yield point, the quality level, and the deoxygenation method, the four parts in order. And "Q" represents the yield point; the value of yield point includes 195 MPa, 215 MPa, 235 MPa, 255 MPa and 275 MPa; the quality level is expressed by the content of sulfur and phosphor: A, B, C and D, in decreasing order; the deoxygenation method is expressed as follows: F represents rimmed steel, b represents semi-killed steel, Z and TZ represents fully-killed steel and special fully-killed steel, and Z and TZ can be omitted in the grades of steel.

For example, Q235 - A · F represents A-grade rimmed steel with the yield point of 235 MPa.

(2) Technical Requirements

The chemical components of each steel grade should accord with Table 2.1. The mechanical properties and technological characteristics should be in line with Table 2.2 and Table 2.3.

Table 2.1 Chemical components of carbon structural steel (GB 700-88)

Grade	Level	Chemical Component/%					Deoxygenation
		C	Mn	Si	S	P	
				≤			
Q195	—	0.06-0.12	0.25-0.50	0.30	0.050	0.045	F, b, Z
Q215	A	0.09-0.15	0.25-0.55	0.30	0.050	0.045	F, b, Z
	B				0.045		
Q235	A	0.14-0.22	0.30-0.65	0.30	0.050	0.045	F, b, Z
	B	0.12-0.20	0.30-0.70		0.045		
	C	≤0.18	0.35-0.80		0.040	0.040	Z
	D	≤0.17			0.035	0.035	TZ
Q255	A	0.18-0.28	0.40-0.70	0.30	0.050	0.045	F, b, Z
	B				0.045		
Q275	—	0.28-0.38	0.50-0.80	0.35	0.050	0.045	b, Z

Table 2.2 Mechanical properties of carbon structural steel (GB 700-88)

Grade	Level	Tensile Test														Impact Test	
		Yield Point σ_s/(N/mm^2)						Tensile Strength σ_b /(N/mm^2)	Elongation δ_s/%							Temperature /°C	V Impact Work (Vertical) /J
		Thickness of Steel (Diameter) /mm							Thickness of Steel (Diameter) /mm								
		≤16	>16-40	>40-60	>60-100	>100-140	>150		≤16	>16-40	>40-60	>60-100	>100-140	>150			
		≤							≤								≥
Q195	-	(195)	(185)	-	-	-	-	315-390	33	32	-	-	-	-		-	-
Q215	A	215	205	195	185	175	165	335-410	31	30	29	28	27	26		-	-
	B															20	27
Q235	A	235	225	215	205	195	185	375-460	26	25	24	23	22	21		-	-
	B															20	27
	C															-	27
	D															-20	27
Q255	A	255	245	235	225	215	205	410-510	24	23	22	21	20	19		-	-
	B															20	27
Q275	-	275	265	255	245	235	225	490-610	20	19	18	17	16	15		-	-

Table 2.3 Technological characteristics of carbon structural steel (GB 700-88)

Grade	Direction of Samples	Cold Bending Test (B=2a, 180°)		
		Thickness of Steel (Diameter)/mm		
		60	>60-100	>100-200
		Diameter of Bending Heart d		
Q195	Vertical	0	-	-
	Horizontal	0.5a		
Q215	Vertical	0.5a	1.5a	2a
	Horizontal	a	2a	2.5a
Q235	Vertical	a	2a	2.5a
	Horizontal	1.5a	2.5a	3a
Q255	-	2a	3a	3.5a
Q275	-	3a	4a	4.5a

Note: B is the width of sample; a is the thickness of steel (diameter).

(3) Selection of Carbon Structural Steel

The selection of steel depends on the quality, properties and the corresponding standards of steel on one side; on the other side, it depends on the requirements of the project for the properties of steel.

In national standards, carbon structural steel includes five grades, and each grade has different quality levels. Generally, the bigger the grade is, the higher the carbon content is, and the toughness is. The martin steel and oxygen converter steel have good quality, and steel grade D and grade C with lower sulfur and phosphor contents are better than steel grade B and grade A.

Of all the steel structures, Q235 is most commonly used for its advantages over others. It has higher strength, better plasticity, toughness and welding, which are also the requirements of the common steel structure. C and D grades of Q235 contain less S P, which makes them important in welding structure. It applies particularly to those structures under impact load at low temperature.

Steel Q215 has low strength and high plasticity, and deforms a lot under stress. It can replace Q235 after cold working.

Steel Q275 has high strength but low plasticity, and sometimes is rolled to ribbed bars used in concrete.

2. Low-alloy High-strength Structural Steel

(1) Representation of Grades

According to *Low-alloy High-strength Structural Steel* (GB/T 1591-94), the national standard, there are five grades. The known elements are manganese, silicon, barium, titanium, niobium, chromium, nickel and lanthanum. The representation of grades consists of the letter of the yield point, the value of the yield point, and the quality level (including A, B, C, D, E, the five levels).

(2) Standards and Properties

Table 2.4 and Table 2.5 show the chemical elements and mechanical properties of the low-alloy high-strength structural steel.

(3) Application

The addition of alloy elements into the steel can modify the organization and properties of steel. If 18Nb or 16Mn (the yield point is 345 MPa) with the similar carbon content (0.14%-0.22%) is compared with Q235 (the yield point is 235 MPa), the yield point is improved by 32%, and it has good plasticity, impact toughness and weld ability and can resist low temperature and corrosion; and under the same conditions, it can make the carbon structural steel save steel consumption by 20%-30%.

The ore or the original alloy elements in steel waste, such as niobium and chromium, are often used for the alloying of steel; or some cheap alloy elements, such as silicon and manganese, are added; if there is special requirement, a little amount of alloy elements, such as titanium and vanadium, can be used. The smelting equipment is basically the same with the equipment to produce carbon steel, so the cost increases a little.

The adoption of low-alloy structural steel will reduce the weight of structures and extend the useful time, and the high-strength low-alloy structural steel is especially used in the large-span or large column-grid structures for better technical and economic effects.

Table 2.4 Chemical components of low-alloy high-strength structural steel (GB/T 1591-94)

Grade	Quality Level	Chemical Components/%										
		C (≤)	Mn	Si	P (≤)	S (≤)	V	Nb	Ti	Al (≥)	Cr (≤)	Ni (≤)
Q295	A	0.16	0.80-1.50	0.55	0.045	0.045	0.02-0.15	0.015-0.060	0.02-0.20	-		
	B	0.16	0.80-1.50	0.55	0.040	0.040	0.02-0.15	0.015-0.060	0.02-0.20	-		
Q345	A	0.02	1.00-1.60	0.55	0.045	0.045	0.02-0.15	0.015-0.060	0.02-0.20	-		
	B	0.02	1.00-1.60	0.55	0.040	0.040	0.02-0.15	0.015-0.060	0.02-0.20	-		
	C	0.20	1.00-1.60	0.55	0.035	0.035	0.02-0.15	0.015-0.060	0.02-0.20	0.015		
	D	0.18	1.00-1.60	0.55	0.030	0.030	0.02-0.15	0.015-0.060	0.02-0.20	0.015		
	E	0.18	1.00-1.60	0.55	0.025	0.025	0.02-0.15	0.015-0.060	0.02-0.20	0.015		
Q390	A	0.20	1.00-1.60	0.55	0.045	0.045	0.02-0.20	0.015-0.060	0.02-0.20	-	0.3	0.7
	B	0.20	1.00-1.60	0.55	0.040	0.040	0.02-0.20	0.015-0.060	0.02-0.20	-	0.3	0.7
	C	0.20	1.00-1.60	0.55	0.035	0.035	0.02-0.20	0.015-0.060	0.02-0.20	0.015	0.3	0.7
	D	0.20	1.00-1.60	0.55	0.030	0.030	0.02-0.20	0.015-0.060	0.02-0.20	0.015	0.3	0.7
	E	0.20	1.00-1.60	0.55	0.025	0.025	0.02-0.20	0.015-0.060	0.02-0.20	0.015	0.3	0.7
Q420	A	0.20	1.00-1.70	0.55	0.045	0.045	0.02-0.20	0.015-0.060	0.02-0.20	-	0.4	0.7
	B	0.20	1.00-1.70	0.55	0.040	0.040	0.02-0.20	0.015-0.060	0.02-0.20	-	0.4	0.7
	C	0.20	1.00-1.70	0.55	0.035	0.035	0.02-0.20	0.015-0.060	0.02-0.20	0.015	0.4	0.7
	D	0.20	1.00-1.70	0.55	0.030	0.030	0.02-0.20	0.015-0.060	0.02-0.20	0.015	0.4	0.7
	E	0.20	1.00-1.70	0.55	0.025	0.025	0.02-0.20	0.015-0.060	0.02-0.20	0.015	0.4	0.7
Q460	C	0.20	1.00-1.70	0.55	0.035	0.035	0.02-0.20	0.015-0.060	0.02-0.20	0.015	0.7	0.7
	D	0.20	1.00-1.70	0.55	0.030	0.030	0.02-0.20	0.015-0.060	0.02-0.20	0.015	0.7	0.7
	E	0.20	1.00-1.70	0.55	0.025	0.025	0.02-0.20	0.015-0.060	0.02-0.20	0.015	0.7	0.7

Note: Al in the table is the total aluminum content. If the acid-soluble aluminum is tested, the content should be no less than 0.010%.

Table 2.5 Mechanical properties of low-alloy high-strength structural steel (GB/T 1591-94)

Grade	Level	Tensile Test										180°Bending Test (d-diameter of bending heart; a-thickness of specimen) (diameter)	
		Yield Point σ_s/MPa Thickness (diameter, side length)/mm				Tensile Strength σ_b/MPa	Elongation δ_s/%	Impact work(A_{kv})(Vertical)/J				Thickness (diameter)/mm	
		≤15	>16-35	>35-50	>50-100			+20 °C	0 °C	-20 °C	-40 °C		
		≥						≥				≤16	>16-100
Q295	A	295	275	255	235	390-570	23	34				d=2a	d=3a
	B	295	275	255	235	390-570	23					d=2a	d=3a
Q345	A	345	325	295	275	470-630	21					d=2a	d=3a
	B	345	325	295	275	470-630	21					d=2a	d=3a
	C	345	325	295	275	470-630	22	34				d=2a	d=3a
	D	345	325	295	275	470-630	22		34	34	27	d=2a	d=3a
	E	345	325	295	275	470-630	22					d=2a	d=3a
Q390	A	390	370	350	330	490-650	19					d=2a	d=3a
	B	390	370	350	330	490-650	19					d=2a	d=3a
	C	390	370	350	330	490-650	20	34				d=2a	d=3a
	D	390	370	350	330	490-650	20		34	34	27	d=2a	d=3a
	E	390	370	350	330	490-650	20					d=2a	d=3a
Q420	A	420	380	380	360	520-680	18					d=2a	d=3a
	B	420	380	380	360	520-680	18					d=2a	d=3a
	C	420	380	380	360	520-680	19	34				d=2a	d=3a
	D	420	380	380	360	520-680	19		34	34	27	d=2a	d=3a
	E	420	380	380	360	520-680	19					d=2a	d=3a
Q460	C	460	420	360	360	550-720	17					d=2a	d=3a
	D	460	420	360	360	550-720	17		34	34	27	d=2a	d=3a
	E	460	420	360	360	550-720	17					d=2a	d=3a

2.3.2 Steel for Concrete Structures

Recently, the steel used for concrete structures mainly includes: hot-rolled reinforced bar, cold-drawn hot-rolled reinforced bar, cold-drawn low-carbon steel wire, cold-rolled ribbed bar, heat-tempering bar, steel wire and strand for pre-stressed concrete, and cold-rolled-twisted bar.

1. Hot-rolled Reinforced Bar

The hot-rolled reinforced bars used for concrete structures should have high strength, a certain plasticity, toughness, cold bending and weld ability. The hot-rolled reinforced bars mainly are the plain round bar rolled by Q 235 and the ribbed steel made of alloy steel.

(1) Standard and Property of Hot-rolled Reinforced Bar

Based on *Hot-rolled Plain Round Steel Bars for the Reinforcement of Concrete* (GB 13013), the national standard, the hot-rolled vertical round bars are level I, and the strength grade is HPB 235(Table 2.6); the grades of the plain steel bars are represented by HRB and the minimum value of the yield point of the grade, and grades include HRB335, HRB400, and HRB500. H represents "hot-rolled", R represents "ribbed", and B represents "bar", the numbers afterwards represents the minimum value of the yield point (Table 2.7).

Table 2.6 Technical requirements for hot-rolled plain round bars

Surface Shape	Bar Level	Strength Grade	Nominal Diameter/mm	Yield Point σ_s /MPa	Tensile Strength σ_b /MPa	Elongation δ_s/%	Cold Bending (d-diameter of bending heart a-nominal diameter of bar)
				≥			
Plain Round	I	HPB235	8-20	235	370	25	180°d=a

Table 2.7 Grades and technical requirements for hot-rolled bars

Grade	Nominal Diameter /mm	σ_s ($\sigma_{p0.2}$)/MPa	σ_b/MPa	δ_s/%
		≥		
HRB335	6-25	335	490	16
	28-50			
HRB400	6-25	400	570	14
	28-50			
HRB500	6-26	500	630	12
	28-51			

(2) Application

Steel bar grade I or HRB 335 and HRB 400 can be used as the non-prestressed bars in ordinary concrete based on the using conditions; the pre-stressed bars should be HRB400 or HRB 335. The hot-rolled bars grade I is the plain round bars, and others are the crescent ribbed bars whose coarse surface can improve the gripping power between concrete and steel bars.

2. Cold-drawn Hot-rolled Bar

Cold-drawn hot-rolled bar is made at the room temperature by drawing the hot-rolled steel bar with a kind of stress up to or beyond the yield point but less than the tensile strength and then unloading.

The cold drawing can improve the yield point by 17%-27%, the material will become brittle, the yield stage becomes short, the elongation decreases, but the strength after cold-drawn ageing will increase a little.

In practice, all the cold drawing, derusting, straightening, and cutting can be combined into one process, which simplifies the procedure and improves the efficiency; cold drawing can save steel and make pre-stressed bars, which increases the varieties of steel, and the equipment is simple and easy to operate, so it is one of the most common method for the cold working of steel. According to *Construction and Acceptance Codes for Concrete Structures* (GB 50204-2002), the national standard, the technical requirements should be in line with Table 2.8.

Table 2.8 Properties of cold-drawn hot-rolled bars (GB 50204-2002)

Grade	Diameter/mm	Yield Strength /(N/mm^2)	Tensile Strength /(N/mm^2)	Elongation δ_{10}/%	Cold Bending	
			≥		Bending Angle	Bending Diameter/mm
HRB235	≤12	280	370	11	180°	3*d*
HRB335	≤25	450	510	10	90°	3*d*
	28-40	430	490	10	90°	3*d*
HRB400	8-40	500	570	8	90°	3*d*
HRB500	10-28	700	835	6	90°	3*d*

Note: ① *d* is the diameter of steel bar(mm);

② The value of the yield strength of cold-drawn bars in the table is the standard strength value of cold-drawn bars regulated in *Design Specification for Concrete Structure*, the existing national standard;

③ The cold bending diameter of the cold-drawn steel bars HRB400 and HRB500 with the diameter more than 25 mm should increase 1*d*.

3. Cold-rolled Ribbed Bar

The cold-rolled ribbed bar is the bar made by cold drawing or cold rolling the ordinary low-carbon steel, the quality carbon steel or the low-alloy hot-rolled coiled bar to reduce the diameter and form crescent cross ribs on three faces or two faces of the bar.

The base metal of the cold-rolled ribbed bar should be in line with the existing national standard *Cold-rolled Ribbed Bar* (GB 13788). At present, most of the cold-rolled ribbed bars produced at home adopt passive cold rolling machine to reduce diameter and form crescent cross ribs on three faces of bars.

The other one is the active rolling machine which can reduce diameter and form crescent cross ribs on two faces of bars.

Cold-rolled ribbed bar uses CRB as the grade code. According to JGJ 95-2003 and J254-2003, the cold-rolled ribbed bar has five grades divided by tensile strength: CRB550, CRB650, CRB800, CRB970, and CRB1170. C represents "cold-rolled", R represents "ribbed", and B represents "bar". The value is the minimum value of tensile strength. The mechanical and technological properties of the cold-rolled ribbed bars should be in line with Table 2.9.

Table 2.9 Mechanical and technological properties of cold-rolled ribbed bars (JGJ 95-2003)

Grade	Tensile Strength σ_b /MPa	Elongation /%		Cold Bending180° d-diameter of bending heart; a-nominal diameter of bar	Alternating bending frequency	Relaxation ratio (initial stress $\sigma_{con}=0.7\sigma_b$)	
		δ_{10}	δ_{100}			1000 h(≤)/%	10 h(≤)/%
CRB550	550	8	-	$d=3a$	-	-	-
CRB650	650	-	4	-	3	8	5
CRB800	800	-	4	-	3	8	5
CRB970	970	500	4	-	3	8	5
CRB1170	1170	-	4	-	3	8	5

Note: ① There should be no crack on the surface of the bending parts.

② If the nominal diameters of the bars are 4 mm, 5 mm and 6 mm, the bending diameter of the alternating bending should be 10 mm, 15 mm, and 15 mm respectively.

③ For various bans supplied in coils, their tensile strength after straightening should be still in line with the table.

④ δ_{10} is the elongation of the bar whose standard measured distance is 10 times of its diameter; δ_{100} is the elongation of the bar whose standard measured distance is 100 mm.

The cold-rolled ribbed steel bars have high strength, good plasticity, high cohesion force with concrete, and stable quality. Grade 550 steel bars are mainly used for reinforced concrete structures, especially the main load bearing bars of slab members and the non-prestressed steel bars in pre-stressed concrete structures.

Based on the need of projects and the actual conditions of materials, the cold-rolled ribbed steel bars with diameter of 4-12 mm can be upgraded by 0.5 mm. When grade 550 steel bars are used as the main load-bearing bars, their diameters should be 5-12 mm.

At present, the diameter of the steel bars used greatly in cast-in reinforced concrete slabs is 6 mm above. Grade 650 bars are mainly used in the pre-stressed hollow slabs, with the diameter of 5 mm or 6 mm in several places. Grade 800 bars are the low-alloy coiled bars with diameter of 6.5 mm and strength of 550 MPa.

4. Heat-tempering Bar

Heat tempering is a technological process that the steel is heated, insulated, and cooled based on some rules to make its organization change and gain a required property. Heat-tempering bar is the bar made by quenching and high tempering the hot-rolled ribbed bar (middle-carbon low-alloy steel). Its plasticity decreases little, but its strength increases a lot, and the comprehensive property is ideal. Table 2.10 shows the mechanical indexes of the national standard GB 4463-84.

Heat-tempering bars are mainly used for the pre-stressed concrete sleepers instead of carbon steel wires. Because they are easy to be made, have stable quality and good anchoring ability, and can save steel, they start to be used in pre-stressed concrete projects.

Table 2.10 Mechanical properties of heat-tempering bars (GB 4463-84)

Nominal Diameter/mm	Grade	Yield Point/MPa (kgf/mm^2)	Tensile Strength/MPa (kgf/mm^2)	Elongation δ_{10}/%
		≥		
6	$40Si_2Mn$	1325(135)	1470(150)	6
8.2	$48Si_2Mn$			
10	$45Si_2Cr$			

5. Cold-drawn Low-carbon Steel Wire

The cold-drawn low-carbon steel wire is made by tungsten alloy wire-drawing model whose cross-section is less than Q235 (or Q215) coiled bars with diameter of 6.5-8.0 mm. The cold-drawn steel wire undertakes not only tension but also extrusion.

The national standard GB 50204-92 regulates that the cold-drawn low-carbon steel wire has two grades of strength: the first grade is pre-stressed wire, and the second grade is non-prestressed wire. When a concrete plant conducts cold-drawing by itself, it should strictly control the quality of steel wires and check their appearances in batches randomly. The plant should check the coiled bars one by one to find whether their mechanical and technical properties are in line with Table 2.11. All the bars whose elongation is unqualified should not be used in the pre-stressed concrete members.

Table 2.11 Mechanical and technological properties of cold-rolled ribbed bars (GB 50204-92)

Grade	Diameter/mm	Tensile Strength/MPa		Elongation δ_{10}/%	180° Repeated Bending (number)
		Group 1	Group 2		
		≥			
First Grade	5 4	650(65) 700(70)	600(60) 650(65)	3 2.5	4
Second Grade	3-5	550(55)		2.0	4

Note: After the pre-stressed cold-drawn low-carbon steel wire is adjusted by machine, the standard tensile strength should be decreased by 50 MPa.

6. Pre-stressed Steel Wire for Concrete or Steel Strain

They are the special products made by cold working, re-backfiring, cold rolling or crossing the high-quality carbon structural steel, also called high-quality carbon steel wire or steel strain.

The national standard GB/T 5223-2002 regulates that the pre-stressed steel wire for concrete can be divided by processing way: cold-drawn steel wire (code of WCD) and stress-relieved wire, the two types. The stress-relieved wire can be divided into low loose plain round wire (code of P), spiral rib steel wire (code of H), and deformed steel wire (code of I), the three types. The mechanical properties of cold-drawn wire, stress-relieved wire, spiral rib steel wire, and stress-relieved deformed wire are shown in Table 2.12, Table 2.13 and Table 2.14.

Table 2.12 Mechanical property of cold-drawing steel wires

Nominal Diameter d_n/mm	Tensile Strength σ_b/MPa (≥)	Specified Non-proportional Elongation Stress $\sigma_{p0.2}$/MPa (≥)	Total Elongation under the Maximum Stress δ/% (≥)	Bending Number (180°) (≥)	Bending Diameter R /mm	Shrinkage Ratio of Section φ/% (≥)	Twisting Number of Every 210 mm Torque	Relaxation Ratio after 1000h, when the initial stress equals to 70% of nominal tensile strength r /% (≤)
3.00	1470	1100	1.5	4	7.5	-	-	8
4.00	1570	1180		4	10	35	8	
5.00	1670	1250		4	15		8	
	1770	1330						
6.00	1470	1100		5	15	30	7	
7.00	1570	1180		5	20		6	
8.00	1670	1250		5	20		5	
	1770	1330						

Table 2.13 Mechanical properties of stress-relieved plain round and spiral rib steel wires

Nominal Diameter d_n /mm	Tensile Strength σ_b/Mpa (≥)	Specified Non-proportional Elongation Stress $\sigma_{p0.2}$/MPa (≥)		Total Elongation under the Maximum Stress δ /% (≥)	Bending Number (180°) (≥)	Bending Diameter R/mm	Stress Relaxation Property		
							Percentage of Initial Stress to Nominal Tensile Strength/%	Relaxation Ratio after 1000 h r/% (≤)	
								WLR	WNR
		WLR	WNR				For All Specifications		
4.00	1470	1290	1250	3.5	3	10			
4.80	1570	1380	1330		4	15			
5.00	1670	1470	1410		4				
	1770	1560	1500		4				
	1860	1640	1580		4				
6.00	1470	1290	1250		4	20	60	1.0	4.5
6.25	1570	1380	1330		4		70	2.0	8
7.00	1670	1470	1410		4		80	4.5	12
	1770	1560	1500		4				
8.00	1470	1290	1250		4				
9.00	1570	1380	1330		4	25			
10.00	1470	1290	1250		4				
12.00					4	30			

Table 2.14 Mechanical properties of stress-relieved deformed wires

<table>
<tr><th rowspan="4">Nominal Diameter d_n/mm</th><th rowspan="4">Tensile Strength σ_b/Mpa (≥)</th><th colspan="2" rowspan="3">Specified Non-proportional Elongation Stress $\sigma_{p0.2}$/MPa (≥)</th><th rowspan="4">Total Elongation under the Maximum Stress δ/% (≥)</th><th rowspan="4">Bending Number (180°) (≥)</th><th rowspan="4">Bending Diameter R/mm</th><th colspan="3">Stress Relaxation Property</th></tr>
<tr><th rowspan="2">Percentage of Initial Stress to Nominal Tensile Strength /%</th><th colspan="2">Relaxation Ratio after 1000 h r/% (≤)</th></tr>
<tr><th>WLR</th><th>WNR</th></tr>
<tr><th>WLR</th><th>WNR</th><th colspan="3">For All Specifications</th></tr>
<tr><td rowspan="5">≤5.0</td><td>1470</td><td>1290</td><td>1250</td><td rowspan="9">3.5</td><td rowspan="9">3</td><td rowspan="5">15</td><td rowspan="9">60
70
80</td><td rowspan="9">1.5
2.5
4.5</td><td rowspan="9">4.5
8
12</td></tr>
<tr><td>1570</td><td>1380</td><td>1330</td></tr>
<tr><td>1670</td><td>1470</td><td>1410</td></tr>
<tr><td>1770</td><td>1560</td><td>1500</td></tr>
<tr><td>1860</td><td>1640</td><td>1580</td></tr>
<tr><td rowspan="4">>5.0</td><td>1470</td><td>1290</td><td>1250</td><td rowspan="4">20</td></tr>
<tr><td>1570</td><td>1380</td><td>1330</td></tr>
<tr><td>1670</td><td>1470</td><td>1410</td></tr>
<tr><td>1770</td><td>1560</td><td>1500</td></tr>
</table>

For the pre-stressed steel wires for concrete, the national standard GB/T 5223-2002 regulates that the mark of the products should contain the following content: pre-stressed steel wire, nominal diameter, tensile strength grade; code of processing state, code of appearance, and standard code.

Steel strand is made by 7 steel wires undertaking crossing hot treatment. The national standard GB 5224-85 regulates that the diameter of steel strand should be 9-15 mm, failure load should be 220 kN, and its yield strength should be 185 kN.

2.4 Corrosion and Protection

When the surface of steel contacts with the surrounding environment under a certain condition, it will be corroded. The corrosion will reduce the load-bearing cross-section of steel, the uneven surface will lead to the convergence of stress, which will lower the load-bearing ability of steel; also, the corrosion will lower the fatigue strength greatly, especially the impact toughness of steel, which will result in the brittle fracture of steel. If the steel bars in concrete are corroded, there will be expansion of volume, which makes the concrete crack along bars. Thus, the measures to resist corrosion should be adopted in order to prevent the corrosion of steel in working.

2.4.1 Corrosion Types and Reasons

1. Chemical Corrosion

Chemical corrosion is a pure chemical corrosion caused by the electrolyte solution or various dry gases (such as O_2, CO_2, SO_2, etc.), without any electric current. Usually, this kind of corrosion will generate loose oxide on the surface of steel by oxidation, and it is very slow under the dry condition, but it will be very fast under high temperature and humidity.

2. Electrochemical Corrosion

When steel contacts with electrolyte solution and generates electric current, there will be the electrochemical corrosion caused by the role of primary battery. The steel contains ferrite, cementite, and non-metal impurities, and all of these components have different electrodes and potentials, which means their activity are diversified; if there is electrolyte, it will be easy to form two poles of primary battery. When the steel contacts with humid media, such as air, water, and earth, a layer of water film will cover its surface and various ions coming from the air dissolves in water, which forms electrolyte. At first, the ferrite in steel lose its electron, that is, $Fe \rightarrow Fe^{2+} + 2e$, to become anode, and cementite becomes cathode. In acidic electrolyte, H^+ obtains electron to become H_2 and runs away; in neutral media, water gets OH^- due to the deoxygenation of oxygen and generates insoluble $Fe(OH)_2$; it can be oxidized into $Fe(OH)_3$ and its dehydration product $Fe_2(OH)_3$ which is the major component for bronze rust.

2.4.2 Corrosion Prevention of Steel

1. Protective Film

This method is to isolate the steel from the surrounding media with the protective film to prevent or delay the damage caused by the corrosion of external corrosive media. For example, paint coatings, enamel or plastic on the surface of steel; or use the metal coating as the protective film, such as zinc, tin, and chrome.

2. Electrochemical Protection

Current-free protection is to connect a piece of metal, such as zinc and magnesium, more active than steel to the steel structure, and because zinc and magnesium have lower potentials than steel, the anodes of the corrosion cells coming from zinc and magnesium have be destroyed, but the steel structure will be protected. This method can be used for the places which are difficult or impossible to be covered with protective layer, such as steam boiler, shell of steamboat, underground pipe, maritime structure, and bridge.

Impressed current protection is to emplace some waste steel or other refractory metals around the steel structure, such as high silicon iron and silver-lead alloy, and to connect the cathode of the impressed direct current to the protected steel structure and the anode to the refractory metals, and the refractory metals become the anode to be corroded and the structure becomes the cathode to be protected.

3. Alloying

The addition of alloy elements into carbon steel to produce various alloy steel will improve its anti-corrosion, such as nickel, chrome, titanium, and copper.

The above method can be adopted to prevent the corrosion of the steel bars in concrete, but the most economic and effective way is to improve the density and the alkalinity of concrete and make sure that the steel bars are thick enough.

In the hydration products of the cement, there is about 1/5 $Ca(OH)_2$, and when the pH value of the media reaches to about 13, there is passive film on the surface of steel bars, so the bars in concrete are difficult to generate rust. But when CO_2 in the air diffuses into the concrete and reacts with $Ca(OH)_2$ to neutralize the concrete. When pH value falls to 11.5 or below, the passive film will be destroyed and the steel surface reveals active state; and if there is humid and oxygen condition, the electrochemical corrosion will start on the surface of steel bars; because the volume of rust is 2-4 times than steel, it will lead to the cracking of concrete along bars. CO_2 diffuses into the concrete and carries the carbonization, so the improvement of the density of concrete will effectively delay the carbonization process.

Because Cl^- will destroy the passive film, the consumption of chloride should be controlled in the preparation of reinforced concrete.

Questions

2.1 What is steel and building steel?

2.2 Describe the application and classification of steel. How many subdivisions of each class?

2.3 What are the main mechanical properties of building steel?

2.4 In the figure of stress-strain curves of low carbon steel, how many stages are there? What are the characteristics and indexes of each stage?

2.5 How many kinds of steel used for concrete structures? What is their application?

2.6 Briefly list the reason of corrosion and the protection measures.

References

[1] HAIMEI ZHANG. Building Materials in Civil Engineering [M]. Beijing: Science Press, 2011.

[2] MICHAEL S MAMLOUK, JOHN P ZANIEWSKI. Materials for Civil and Construction Engineers [M]. Stockton: QWE Press, 2011.

Chapter 3
Aggregate

3.1 Introduction

In civil engineering, the term aggregate means a mass of crushed stone, gravel, sand, etc., predominantly composed of individual particles, but in some cases including clays and silts. The largest particle size in aggregates may have a diameter as large as 150 mm (6 in), and the smallest particle can be as fine as 5 to 10 microns. Aggregates play an important role in highway construction. Without aggregate, concrete bridges and structures, Hot Mix Asphalt (HMA) pavements, and concrete pavements could not be constructed and very few roads would sustain the current loads. The use of aggregates in highway construction has literally brought the transportation industry out of the mud.

3.1.1 Natural Aggregates and Artificial Aggregates

The aggregate is divided into two types according to its different source: natural and aggregates.

Natural aggregates are obtained from natural deposit of sand and gravel or from quarries by cutting rocks. Cheapest among them will be the natural sand and gravel which have been reduced to their present size by natural agents such as water, wind and snow. River deposits are the most common and have good quality.

The second most commonly used source of aggregates is quarried bed rock material. Crushed aggregates are made by breaking down natural bed rocks into requisite graded particles through a series of blasting, crushing, screening, etc.

Artificial aggregates are manufactured aggregates or by-products of industrial processes. Artificial aggregates can use slag waste from iron and steel mills and expanded shale and clays to produce lightweight aggregates. Heavyweight concrete, used for radiation shields, can use steel slag and bearings for the aggregate. Styrofoam beads can be used as an aggregate in lightweight concrete used for insulation.

3.1.2 Geological Classification

All natural aggregates result from the breakdown of large rock masses. Classify rocks into three basic types: igneous, sedimentary, and metamorphic.

Igneous rocks form when molten (hot, liquid) elements cool and solidify into crystals. The crystals that form are rich in the elements silicon and oxygen, by far the most abundant elements comprising Earth's crust and mantle. When bonded together, silicon and oxygen form the silicate ion, the building block for almost all igneous minerals.

Due to the rate of cooling and composition of the constituent materials, there are three main types of igneous rocks: (1) plutonic rocks; (2) volcanic (extrusive) rocks; (3) hypabyssal rocks.

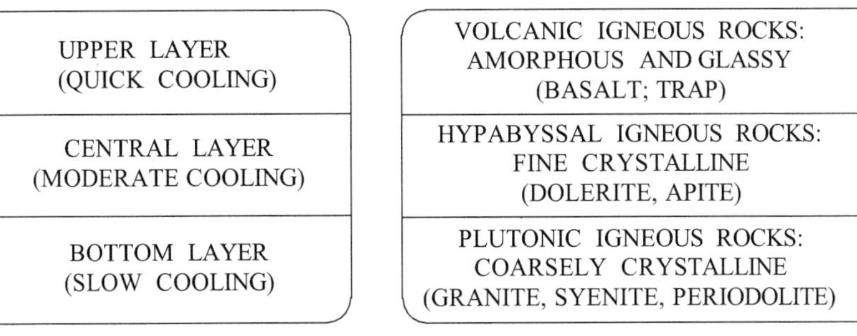

Figure 3.1 Formation of igneous rocks

(1) Plutonic rocks are crystallized from magma slowly cooling at a considerable depth from the surface of the earth. Consequently, such rocks possess coarsely crystalline structure and the size of the crystal is normally large.

The most common rock types in plutons are granite, granodiorite, tonalite, monzonite, and quartz diorite.

(2) Hypabyssal rocks are intrusive igneous rocks that originate at medium to shallow depths within the crust, and has intermediate grain size and often porphyritic texture between that of volcanic and plutonic rocks. Common examples of such rocks are dolerite, dolomite, spinel, etc.

(3) Volcanic rocks are formed from magma erupted from a volcano. The rate of cooling is thus very quick, due to which these rocks are non-crystalline, amorphous, and glassy in texture. The top layer of such rock is quite smooth and shiny. In civil Engineering, the common examples are basalt, rhyolite, andesite, trap, etc.

Sedimentary rocks are types of rock that are formed by the deposition of material at the Earth's surface and within bodies of water. Igneous rocks, formed on the surface of the earth by the cooling of magma, is subjected to natural weathering agents break or disintegrate the surface rocks. These products of weathered rock are then transported by other agents (such as wind) from its place of origin and deposited elsewhere. This process of deposition is called the sedimentation process and the rocks so formed are known as sedimentary rocks. Compared with igneous rocks, sedimentary rocks have low apparent dense and compactness, high water absorption, low strength and poor durability. The example of such rocks are slate, sand stone, lime stone, gypsum, lignite, etc. Limestone is the most widely used in civil engineering. It has been used directly in buildings as load bearing walls and also in facades. Crushed limestone, also called aggregate, is used as a filler in concrete, as a base in road construction, and as a filler in asphalt.

Metamorphic rocks are created by the physical or chemical alteration by heat and pressure of an existing igneous or sedimentary material into a denser form. Due to the action of plate tectonics, compression, stress and shearing forces over long periods of time, rocks can be essentially warped and deformed, causing them to be compacted into a smaller volume of space. As a consequence, metamorphic rocks are always denser than their original material, and also much less susceptible to erosional breakdown. Examples of metamorphic rocks are schist (converted basalt), quartzite (compressed sandstone), and marble (compressed limestone or dolomite).

3.1.3 Course Aggregates and Fine Aggregates

Due to its different size of aggregate particles, there are two types of aggregates.

Course aggregates: aggregate particles that are retained on a 4.75 mm sieve (No. 4). A 4.75 mm sieve has openings equal to 4.75 mm between the sieve wires. A No. 4 sieve has four openings per linear inch. The 4.75 mm sieve is the metric equivalent to a No. 4 sieve.

They can either be from Primary, Secondary or Recycled sources. Primary or "virgin" aggregates are either land-won or marine-won. Gravel is a coarse marine-won aggregate; land-won coarse aggregates include gravel and crushed rock. Gravels constitute the majority of coarse aggregate used in concrete with crushed stone making up most of the remainder.

Secondary aggregates are materials which are the by-products of extractive operations and are derived from a very wide range of materials.

Recycled concrete is a viable source of aggregate and has been satisfactorily used in granular subbases, soil-cement, and in new concrete. Recycled aggregates are classified in two ways:

(1) Recycled Aggregate (RA).

(2) Recycled Concrete Aggregate (RCA).

Fine aggregates: aggregate particles that pass a 4.75 mm sieve (No. 4). As with coarse aggregates these can be from Primary, Secondary or Recycled sources.

3.2 Aggregate Properties

Aggregate particles have certain physical and chemical properties which make the aggregate acceptable or unacceptable for specific uses and conditions.

3.2.1 Physical Properties

The physical properties of aggregates stem from the inherent properties of the source rock, which include texture, structure, and mineral composition.

1. Apparent Density

Apparent density of aggregates is related to the mineral composition and air voids. The apparent density of dense aggregates, such as granite and marble, is 2500-3100 kg/m^3, close to its density. The apparent dense of aggregates with higher air voids is 500-1700 kg/m^3, such as volcanic tuff and pumice.

The most common classification of aggregates on the basis of bulk specific gravity is lightweight, normal-weight, and heavyweight aggregates. In normal concrete the aggregate weighs 1520 -1680 kg/m^3, but occasionally designs require either lightweight or heavyweight concrete. Lightweight concrete contains aggregate that is natural or synthetic which weighs less than 1100 kg/m^3 and heavyweight concrete contains aggregates that are natural or synthetic which weigh more than 2080 kg/m^3.

Pumice is widely used to make lightweight concrete, insulate low-density cinder blocks and wall materials.

2. Water Absorption

Water absorption is related to air voids and pore characteristics. Igneous rocks and some metamorphic rocks has very low value of porosity, consequently, the water absorption is low. The consolidation and compactness of sedimentary rocks is variable due to different forming condition, which results in different water absorption. For example, the water absorption rate of dense limestone is lower than 1%, but that of porous shell limestone is higher than 15%.

The water absorption has big effect on its strength and water resistance.

3. Water Resistance

When there are a lot of clay or soluble substance in the aggregates, the softening coefficient is small and the water resistance is poor. The softening coefficient is used to represent water resistance of aggregate. Due to the different softening coefficient, the aggregates are classified into three levels.

(1) High water-resistant aggregates: softening coefficient is higher than 0.90.

(2) Medium water-resistant aggregates: softening coefficient between 0.7-0.9.

(3) Low water-resistant aggregates: softening coefficient is lower than 0.7.

Generally, the aggregates whose softening coefficient is lower than 0.6 cannot be used in important structure.

4. Frost Resistance

The frost resistance of aggregates is characterized by the number of freeze-thaw cycles. It is permitted to evaluate the frost resistance of aggregates by the number of saturation cycles in a solution of sodium sulfate, with further drying. In terms of frost resistance aggregates are divided into grades: F15, F25, F50, F100, F150, F200, F300, F400. The higher the grade, the higher is the frost resistance of aggregates.

5. Heat Resistance

The heat resistance of aggregates depends on its chemical components and mineral composition. The aggregates that contains gypsum begin to break when the temperature exceeds 100℃. The aggregates that contains magnesium carbonate and calcium carbonate damages above 625℃ and 827℃, respectively. The strength of some aggregates, such as granite, consisting of quartz and some other minerals will drop sharply when quartz expand with heat.

6. Thermal Conductivity

The thermal conductivity of aggregates is related to apparent density and structure. It can reach 2.91-3.49 W/(m · K). The thermal conductivity of aggregates with the same composition in glassy state is lower than that in crystalline state.

3.2.2 Mechanical Properties

1. Compressive Strength

According to *Code for Design of Masonry Structures* (GB 50003-2001), cube compressive strength of 70 mm × 70 mm × 70 mm is represented as the compressive strength of aggregates for masonry. The strength is divided into seven grades, which are MU100, MU80, MU60, MU50, MU40, MU30 and MU20. According to *Test Methods for Natural Facing Stones* (GB 9966-2001), compressive strength of cube with size of 50 mm × 50 mm × 50 mm or cylinder with a diameter of 50 mm and a height of 50 mm is taken as the compressive strength of facing stones. The strength of sedimentary rock is related to colloidal substances, the compressive strength of which is higher when siliceous materials are main colloids, followed by calcareous materials, and then argillaceous materials.

The compressive strength is decided by mineral composition, structure, types of colloidal substances and homogeneity, etc. For example, the quartz is the main mineral substance in granite. It makes granite has high strength for its high stiffness. Mica is a kind of flaky mineral, which is subject to splitting into flakes. The higher the content of mica, the lower the strength of aggregates is.

2. Impact Strength

Aggregates are typical brittle materials, whose impact strength is much less than compressive strength. Mineral composition and structure decide the impact strength. In general, the impact strength of aggregates with crystal structure is higher than that with amorphous structure. For example, quartz rock and siliceous rock have high brittleness. In comparison with that, gabbro and diabase have good toughness.

3. Hardness

The hardness of aggregates is represented by Mohs hardness or Shore hardness, which is dependent on the hardness of mineral composition and structure. Generally, the higher the compressive strength is, the higher the hardness is. The aggregates with higher hardness have better abrasion and scratch resistance but poor surface processing performance.

4. Abrasion Resistance

Abrasion resistance refers to the ability of aggregates to resist abrasion, shear and impact in service. It is represented by abrasion loss per unit area. The abrasion resistance of aggregates has to do with the hardness of mineral substance, structure, compressive strength, impact strength and so on. As for natural stone, abrasion value is obtained according to *Test Methods for Aggregate for Highway Engineering-Part 4: Abrasion Resistance of coarse aggregate* (JTG E42-2005). As for coarse aggregates in concrete, the Los Angeles (L.A.) abrasion test is a common test method used to indicate aggregate abrasion characteristics.

3.2.3 Technology Characteristic

The technology characteristic of aggregates refers to the difficult and possibility in aggregate mining and processing operation, which include processing performance, polishability and drillability.

1. Processing Performance

Processing performance is an ability to quarry, split, cut, chisel, grind and polish the rocks. It's difficult to process the rocks with high strength, stiffness and toughness. Rocks that are brittle, rough or have particles staggered or layer structure and weathered rocks, cannot meet the processing requirement.

2. Polishability

Polishability is the property of rocks being polished and smoothed. Rocks containing compact and uniform structure and fine particles have good polishability, which can be polished to smooth surface. Rocks containing loose, porosity and flaky structure have poor polishability.

3. Drillability

Drillability refers to the difficult level of drilling the rocks, which is affected by a lot of complicated factors. In general, it has to do with the strength, stiffness and some other properties. The higher of the strength and stiffness of rocks are, the more difficult it is to be drilled.

3.3 Handling Aggregates

In order to control the quality of the aggregates, aggregates must be handled and stockpiled in such a way as to minimize segregation, degradation, and contamination.

The drop height should be limited to avoid breakage, especially for large aggregates.

Vibration and jiggling on a conveyor belt tends to work fine material downward while coarse particles rise. Segregation can be minimized by moving the material on the belt frequently (up and down, side to side, in and out) or by installing a baffle plate, rubber sleeve, or paddle wheel at the end of the belt to remix coarse and fine particles. Rounded aggregates segregate more than crushed aggregates. Meanwhile, large aggregates segregate more readily than smaller aggregates. Therefore, different sizes should be stockpiled and batched separately. Stockpiles should be separated by dividers or placed in bins to avoid mixing and contamination.

There are some guidelines on recommended practices for handling aggregates, which includes three broad areas: hauling to stockpile, stockpiling, unloading stockpiles and charging mixer.

3.3.1 Hauling to Stockpile

The hauling of aggregates and sand from the point of production is usually by truck or rail; however, there are many circumstances where economics dictate that other modes or combinations be used.

Regardless of mode used, the possibilities of degradation, segregation and contamination arise each time the material is shipped.

(1) Economics is the most important factor to consider when selecting the route and method of transportation.

(2) Care must be exercised through all stages of transportation to avoid segregation of particle sizes and in some cases degradation.

3.3.2 Stockpiling

The method of stockpiling has a large effect on the variation in gradation and degradation characteristics of an aggregate. The most economical and acceptable method of forming and reclaiming stockpiles from aggregates delivered in trucks is to discharge the loads in such a way that they are rightly joined and to reclaim the aggregate with a front-end loader. When aggregates are not delivered in trucks, the least expensive acceptable results are obtained by forming the stockpiles in layers with a crane bucket and reclaiming the aggregates with a front-end loader.

Smaller size aggregate pieces have greater surface area than larger ones. Consequently, when undersized particles are concentrated, through segregation or degradation, water demand is significantly increased.

3.3.3 Unloading Stockpiles and Charging Mixer

Special attention should be given to unloading stockpiles and blending of materials on the charging belt in order to assure uniformity and homogeneity of the concrete produced.

Research has demonstrated that segregation of the mix can result if proper blending of the materials is not accomplished prior to entry into the mixer.

Questions

3.1 According to different reasons, how to classify the aggregate?

3.2 What physical properties does aggregate contain?

3.3 How to text the compressive strength of aggregate?

3.4 Why aggregates of different sizes should be stockpiled and batched separately?

3.5 What is the influences of the method of stockpiling?

References

[1] MICHAEL S MAMLOUK, JOHN P ZANIEWSKI. Materials for Civil and Construction Engineers [M]. Stockton: QWE Press, 2011.

[2] B C PUNMIA, ASHOK KUMAR JAIN, ARUN KR JAIN. Basic Civil Engineering [M]. Firewall Media, 2003.

Chapter 4

Cement

4.1 Introduction

Portland cement is the most common type of cement in general use around the world as a basic ingredient of concrete, mortar, stucco, and non-specialty grout. It was developed from other types of hydraulic lime in England in the early 19th century by Joseph Aspdin, and usually originates from limestone. It is a fine powder, produced by heating limestone and clay minerals in a kiln to form clinker, grinding the clinker, and adding 2 to 3 percent of gypsum. Several types of Portland cement are available. The most common, called ordinary Portland cement (OPC), is grey, but white Portland cement is also available. Its name is derived from its resemblance to Portland stone which was quarried on the Isle of Portland in Dorset, England. It was named by Joseph Aspdin who obtained a patent for it in 1824. However, his son William Aspdin is regarded as the inventor of "modern" Portland cement due to his developments in the 1840s.

The low cost and widespread availability of the limestone, shales, and other naturally-occurring materials used in Portland cement make it one of the lowest-cost materials widely used over the last century. Concrete produced from Portland cement is one of the world's most versatile construction materials.

Portland cement was developed from natural cements made in Britain beginning in the middle of the 18th century. Its name is derived from its similarity to Portland stone, a type of building stone quarried on the Isle of Portland in Dorset, England.

The development of modern Portland cement (sometimes called ordinary or normal Portland cement) began in 1756, when John Smeaton experimented with combinations of different limestones and additives, including trass and pozzolana, relating to the planned construction of a lighthouse, now known as Smeaton's Tower. In the late 18th century, Roman cement was developed and patented in 1796 by James Parker. Roman cement quickly became popular, but was largely replaced by Portland cement in the 1850s. In 1811, James Frost produced a cement he called British cement. James Frost is reported to have erected a manufactory for making of an artificial cement in 1826. In 1811 Edgar Dobbs of Southwark patented a cement of the kind invented 7 years later by the French engineer Louis Vicat. Vicat's cement is an artificial hydraulic lime, and is considered the "principal forerunner" of Portland cement.

The name Portland cement is recorded in a directory published in 1823 being associated with a William Lockwood and possibly others①. In his 1824 cement patent, Joseph Aspdin called his invention "Portland cement" because of its resemblance to Portland stone. However, Aspdin's cement was nothing like modern Portland cement, but was a first step in the development of modern Portland cement, and has been called a "proto-Portland cement".

William Aspdin had left his father's company, to form his own cement manufacture. In the 1840s, William Aspdin accidentally produced calcium silicates which are a middle step in the development of Portland cement. In 1848, William Aspdin further improved his cement. Then, in 1853, he moved to Germany, where he was involved in cement making. William Aspdin made what could be called "meso-Portland cement" (a mix of Portland cement and hydraulic lime). Isaac Charles Johnson further refined the production of "meso-Portland cement" (middle stage of development), and claimed to be the real father of Portland cement.

In 1859, John Grant of the Metropolitan Board of Works, set out requirements for cement to be used in the London sewer project. This became a specification for Portland cement. The next development in the manufacture of Portland cement was the introduction of the rotary kiln, patented by Frederick Ransome in 1885 (U.K.) and 1886 (U.S.); which allowed a stronger, more homogeneous mixture and a continuous manufacturing process. The Hoffmann "endless" kiln which was said to give "perfect control over combustion" was tested in 1860, and showed the process produced a better grade of cement. This cement was made at the Portland Cementfabrik Stern at Stettin, which was the first to use a Hoffmann kiln. The Association of German Cement Manufacturers issued a standard on Portland cement in 1878.

Portland cement had been imported into the United States from Germany and England, and in the 1870s and 1880s, it was being produced by Eagle Portland cement near Kalamazoo, Michigan. In 1875, the first Portland cement was produced in the Coplay Cement Company Kilns under the direction of David O. Saylor in Coplay, Pennsylvania. By the early 20th century, American-made Portland cement had displaced most of the imported Portland cement.

4.2 Production Process of Portland Cement

Portland cement is a fine powder, gray or white in color, that consists of a mixture of hydraulic cement materials comprising primarily calcium silicates, aluminates and ferrous aluminates. More than 30 raw materials are known to be used in the manufacture of Portland cement, and these materials can be divided into four distinct categories: calcareous, siliceous,

① Doding. "125 Years of Research for Quality and Progress" [OL]. [2002-09-01] https://www.docin.com/p-1867361370.html.

argillaceous, and iron. These materials are chemically combined through high temperature processing and subjected to subsequent mechanical processing operations to form gray and white Portland cement. Gray Portland cement is used for structural applications and is the more common type of cement produced. White Portland cement has lower iron and manganese contents than gray Portland cement and is used primarily for decorative purposes. Portland cement manufacturing plants are part of hydraulic cement manufacturing, which also includes natural, masonry, and pozzolanic cement.

4.2.1 Process Flow of Portland Cement

Portland cement accounts for 95 percent of the hydraulic cement production in the United States. The balance of domestic cement production is primarily masonry cement. Both of these materials are produced in Portland cement manufacturing plants. As shown in the Figure 4.1, the process can be divided into the following primary components: raw materials acquisition and handling, kiln feed preparation, high temperature processing, and finished cement grinding. Each of these process components is described briefly below. The primary focus of this discussion is on high temperature processing operations, which constitute the core of a Portland cement plant.

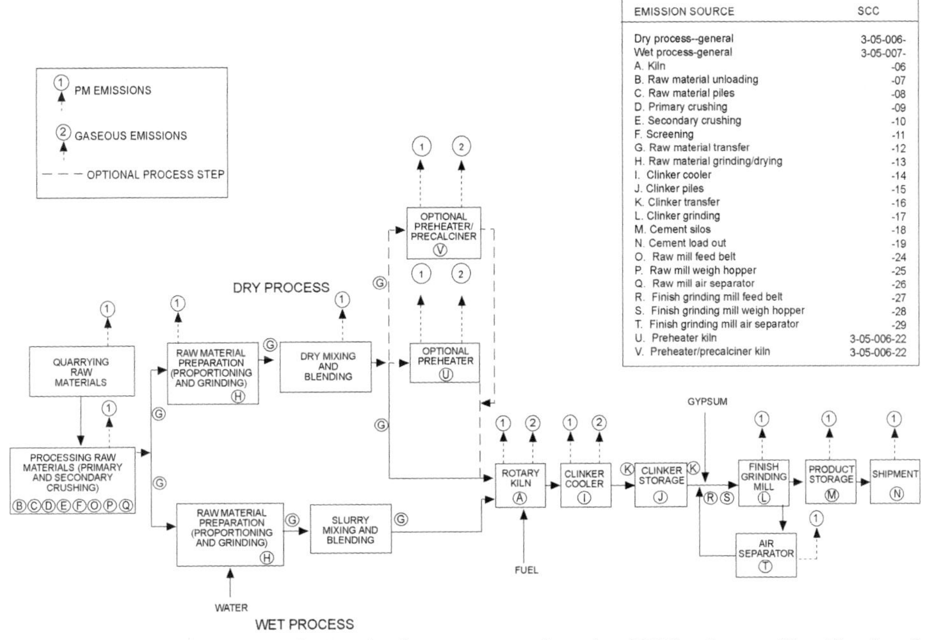

Figure 4.1 Process flow diagram for Portland cement manufacturing (SCC = Source Classification Code)

4.2.2 Portland Cement Production Process

The initial production step in Portland cement manufacturing is raw materials acquisition. Calcium, the element of highest concentration in Portland cement, is obtained from a variety of calcareous raw materials, including limestone, chalk, marl, sea shells, aragonite, and an impure limestone known as "natural cement rock". Typically, these raw materials are obtained from open-face quarries, but underground mines or dredging operations are also gained. Raw materials vary from facility to facility. Some quarries produce relatively pure limestone that requires the use of additional raw materials to provide the correct chemical blend in the raw mix. In other quarries, all or part of the noncalcareous constituents are found naturally in the limestone. Occasionally, pockets of pyrite, which can significantly increase emissions of sulfur dioxide (SO_2), are found in deposits of limestone, clays, and shales used as raw materials for Portland cement. Because a large fraction (approximately one third) of the mass of this primary material is lost as carbon dioxide (CO_2) in the kiln, Portland cement plants are located close to a calcareous raw material source whenever possible. Other elements included in the raw mix are silicon, aluminum, and iron. These materials are obtained from ores and minerals such as sand, shale, clay, and iron ore. Again, these materials are most commonly from open-pit quarries or mines, but they may be dredged or excavated from underwater deposits.

The second step in Portland cement manufacture is preparing the raw mix, or kiln feed, for the high temperature processing operation. Raw material preparation includes a variety of blending and sizing operations that are designed to provide a feed with appropriate chemical and physical properties. The raw material processing operations differ somewhat for wet and dry processes.

Cement raw materials are received with an initial moisture content varying from 1 to more than 50 percent. If the facility uses dry process kilns, this moisture is usually reduced to less than 1 percent before or during grinding. Drying alone can be accomplished in impact dryers, drum dryers, paddle-equipped rapid dryers, air separators, or autogenous mills. However, drying can also be accomplished during grinding in ball-and-tube mills or roller mills. While thermal energy for drying can be supplied by exhaust gases from separate, direct-fired coal, oil, or gas burners, the most efficient and widely used source of heat for drying is the hot exit gases from the high temperature processing system.

The final step in Portland cement manufacturing involves a sequence of blending and grinding operations that transforms clinker to finished Portland cement. Up to 5 percent gypsum or natural anhydrite is added to the clinker during grinding to control the cement setting time, and other specialty chemicals are added as needed to impart specific product properties. This finish milling is accomplished almost exclusively in ball or tube mills.

Typically, finishing is conducted in a closed circuit system, with product sizing by air separation.

4.3 Chemical Composition of Cement

4.3.1 Main Components

The raw materials used for the manufacture of cement consist mainly of lime, silica, alumina and iron oxide. These oxides interact with one another in the kiln at high temperature to form more complex compounds. The relative proportions of these oxide compositions are responsible for influencing the various properties of cement, in addition to rate of cooling and fineness of grinding. Table 4.1 shows the approximate oxide composition limits of ordinary Portland cement.

Table 4.1 Approximate oxide composition limits of ordinary portland cement

Oxide	Percent content
CaO	60-67
SiO_2	17-25
Al_2O_3	3.0-8.0
Fe_2O_3	0.5-6.0
MgO	0.1-4.0
Alkalis(K_2O, Na_2O)	0.4-1.3
SO_3	1.3-3.0

4.3.2 Bogue's Compounds

The identification of the major compounds of cement is largely based on Bogue's equations and hence it is called "Bogue's Compounds". The four compounds usually regarded as major compounds are listed in Table 4.2.

Table 4.2 Major compounds of cement

Name of Compound	Formula	Abbreviated Formula
Tricalcium silicate	$3CaO \cdot SiO_2$	C_3S
Dicalcium silicate	$2CaO \cdot SiO_2$	C_2S
Tricalcium aluminate	$3Cao \cdot Al_2O_3$	C_3A
Tetracalcium ferric aluminate	$4CaO \cdot Al_2O_3 \cdot Fe_2O_3$	C_4AF

It is to be noted that for simplicity's sake abbreviated notations are used. C stands for CaO, S stands for SiO_2, A for Al_2O_3, F for Fe_2O_3 and H for H_2O. The equations suggested by Bogue for calculating the percentages of major compounds are given below.

$$C_3S=4.07(CaO) - 7.60(SiO_2) - 6.72(Al_2O_3) - 1.43(Fe_2O_3) - 2.85(SO_3)$$

$$C_2S=2.87(SiO_2) - 0.754(3CaO \cdot SiO_2)$$

$$C_4A=2.65(Al_2O_3) - 1.69(Fe_2O_3)$$

$$C_4AF=3.04(Fe_2O_3)$$

4.3.3 Minor Compounds

1. K_2O and Na_2O

In addition to the four major compounds, there are many minor compounds formed in the kiln. Two of the minor oxides namely K_2O and Na_2O referred to as alkalis in cement are of some importance and Expressed in terms of Na_2O. These alkalis basically react with active silica in aggregate and produce what is called alkali-silica gel of unlimited swelling type under favorable conditions of moisture and temperature in voids and cracks and further it causes disruption and pattern cracking.

2. Tricalcium Silicate and Dicalcium Silicate

Tricalcium silicate and dicalcium silicate are the most important compounds responsible for strength. Together they constitute 70 to 80 percent of cement. The average C_3S content in modern cement is about 45 percent and that of C_2S is about 25 percent.

The calculated quantity of the compounds in cement varies greatly even for a relatively small change in the oxide composition of the raw materials. To manufacture a cement of stipulated compound composition, it becomes absolutely necessary to closely control the oxide composition of the raw materials.

3. SO_3

SO_3 also appear in cement analysis which comes from adding gypsum 4%-6% during clinker gridding. The Iraqi and British specification for normal high rapid Portland cement pointed that SO_3 content must be between 2.5%-3% according to type of cement and C_3A content.

4. MgO

The percentage of MgO in cement which is come from Magnesia compounds in raw material is about 4% and 5% as maximum range to control expansion from hydration of this oxide in hard concrete.

5. Free Lime

An increase in lime CaO content beyond a certain value makes it difficult to combine with other compounds and free lime will exist in the clinker which causes unsoundness in cement.

6. Insoluble Residue

Insoluble residue is that part of the Cement non-soluble in hydrochloric acid HCl and arise mainly from non-active silica to form cement compounds dissolved in this acid therefore it expresses the completeness of the chemical reactions inside the rotary kiln.

4.4 Setting and Hardening of Cement

4.4.1 Process and Reason of Cement Setting and Hardening

Setting process: the process of losing plasticity with time and becoming dense, which is formed by mixing the cement and water. It happened in two stage initial setting and final setting.

Hardening process: the process of forming hardening cement paste, which loses plasticity and its strength increases with time.

Factors such as fineness and types of cement, blend of gypsum, age, water content, temperature and humidity which influence setting and hardening of Portland cement.

Setting is important in concrete work to keep fresh concrete plastic for enough time which helps the processes of transporting, casting, and compaction.

There are four stage occurs during setting of cement. The first stage: when mixing cement with water, a rapid heat evolution, lasting a few minutes. This heat evolution is probably due to the reaction of cement compounds then, heat evolution ceases quickly. The second stage: this stage called "dormant period" and last 1-4 hours. Also cement particles start to form initial layer of hydration product. Bleeding and sedimentation appear in this stage. The third stage: next heat evolution is on account of dissolution of weak gel which formed first on surfaces of C_3S crystals so water will reach these surfaces and starts to form new gel. This stage last about 6 hours. The forth stage: in this stage cement begin to harden and gain strength.

4.4.2 False Setting

False setting is the rapid development of rigidity in freshly mixed Portland cement paste, mortar, or concrete occurs after few minutes of mixing cement with water without the generation of much heat. It can be controlled by re-mixing without additional water or strength loss. Causes of false setting are:

(1) Drying of gypsum: when gypsum ($CaSO_4 \cdot 2H_2O$) gridding with hot clinker it loss

75% of its water and formed ($CaSO_4 \cdot 1/2\ H_2O$) and if temperature of clinker increase gypsum loss all water in its composition and formed $CaSO_4$. When water add to cement $CaSO_4$ react causes rapid setting.

(2) Bad storage: alkalis in cement react with carbon dioxide forming Alkali carbonate which reacts with calcium hydroxide forming $CaCO_3$ causing setting of cement paste.

(3) Activate effective C_3A exposed to humidity: during bad storage water adhere on cement particles surface and during mixing these active surface combined rapidly with water within minutes.

4.4.3 Flash Setting

Flash setting is a rapid development of rigidity in freshly mixed Portland-cement paste, mortar, or concrete. Further mixing can't dispel this rigidity, and a large amount of heat is produced in the process. It happened due to rapid reaction of aluminates – when insufficient sulfate persent so to prevent it a small amount of gypsum ($CaSO_4 \cdot 2H_2O$) gridding with the cement clinker retards the hydration reaction of tricalcium aluminate so that the calcium silicates can set firstly.

4.5 Soundness of Cement

It refers to the ability of a hardened cement paste to retain its volume after setting without delayed destructive expansion. This destructive expansion is caused by excessive amounts of free lime (CaO) or magnesia (MgO).

The hydrating speed of the over burnt CaO or MgO is slow, CaO or MgO begins to hydrate after cement hardening and causes the hardened cement to expand and crack.

When added too much, gypsum continues to react with calcium aluminate hydrate to form calcium sulphoaluminate hydrate, whose volume increase 1.5 times as big as gypsum and causes the hardened cement paste to crack. At this time, sulphoaluminate hydrate is called cement bacillus.

4.6 Common Cement Types

4.6.1 Ordinary Portland Cement (P.O)

Ordinary Portland cement is the hydraulic cementing material, which is made by grinding a few mixing materials and proper gypsum. The maximum quantity of mixing materials should be no less than 15%.

The sieve residue of ordinary Portland cement 10%, it can be classified into three strength grades (32.5, 42.5, 52.5). Final setting time cannot exceed 10 h. Requirements of strength are in the following Table 4.3.

Table 4.3 Strength demand of ordinary Portland cement

Strength grade	Compressive strength /MPa		Bending strength/MPa	
	3 d	28 d	3 d	28 d
32.5	11.0	32.5	2.5	5.5
32.5R	16.0	32.5	3.5	5.5
42.5	16.0	42.5	3.5	6.5
42.5R	21.0	42.5	4.0	6.5
52.5	22.0	52.5	4.0	7.0
52.5R	26.0	52.5	5.0	7.0

Other requirements of technical properties and range of application are same with Portland cement. The mineral compound is in the range of Portland cement for a few quantity of mixing materials.

The main type of common cement and mainly used in the project of concrete and reinforced concrete. The application range is the same with Portland cement.

4.6.2 Portland Blast-furnace Slag Cement (P.S)

Portland blast-furnace slag cement refers to the cement which is formed by cement clinker, granulated blast-furnace slag and a little gypsum. Quantities of granulated blast-furnace slag is 20%-70%. Its property is different from Portland cement and ordinary cement because Portland blast-furnace slag cement contains less clinker. It can be divided into 6 types: 32.5, 32.5R, 42.5, 42.5R, 52.5, 52. 5R as shown in Table 4.4.

Table 4.4 Strength demands of P.S, P.P, P.F

Strength grade	Compressive strength/MPa		Bending strength/MPa	
	3 d	28 d	3 d	28 d
32.5	10.0	32.5	2.5	5.5
32.5R	15.0	32.5	3.5	5.5
42.5	15.0	42.5	3.5	6.5
42.5R	19.0	42.5	4.0	6.5
52.5	21.0	52.5	4.0	7.0
52.5R	23.0	52.5	4.5	7.0

Characteristics of Portland blast-furnace slag cement, comparing to Portland cement and ordinary Portland cement.

(1) Strength is Low in the Early Days and High in the Later Days

The hydration of slag cement takes 2 steps: hydration of cement clinker and secondary hydration reaction.

In the early days of a secondary hydration, the strength is low with low speed and few hydrations. But it is contrary to the later days.

It is applicable to the prefabricated member cured in the middle days and projects with bearing slowly. But it is not applicable to the projects with the requirement for higher early hardening.

(2) Strong Anti-corrosion and Poor Carbonation Resistance

It is applicable to the water concrete and marine concrete. It can't be used in house with high concentration of CO_2, because of the low content of calcium hydroxide.

(3) Low Hydration Heat

It is applicable to the project with mass concrete.

(4) Temperature Sensitivity

It is applicable to the high temperature cuing because of temperature sensitivity.

(5) Low Anti-freezing

The early hardening is low. Water requirement of volcanic ash is big.

(6) Heat Resistance

It is applicable to the project with requirement of heat resistance concrete.

(7) Poor Permeability

The large quality of mixing materials resulted in sharp arris of blast-furnace slag and low water requirement and retention; therefore, it can form bleeding channels easily.

4.6.3 Portland Pozzolan Cement (P.P)

Portland pozzolan cement refers to the hydraulic cementing material which is produced by Portland cement clinker, Portland pozzolan mixing materials and a little gypsum grinding. Portland pozzolan cement shows excellent impermeability when cured in humid condition or in water. Volcanic ash reacts with lime to have the effect of expansion and forms calcium silicate hydrate, which densifies the structure of hardening cement paste. In addition, Portland pozzolan cement can easily shrink powdery when it is in the dry condition. Because hydration cannot happen in the dry condition and strength keeps stable; calcium silicate hydrate gel can also make cement shrink and form crack. It is not available to the construction in dry area.

4.6.4 Portland Fly-ash Cement (P.F)

Portland fly-ash cement refers to the fly ash and a little gypsum. Quantities of fly ash is 20%-40%. Portland fly-ash cement shows low early-strength. The fly ash forms a double cementing action after 1-3 months because the fly ash surface structure is dense. It is available to the slow construction. In addition, Portland fly-ash cement expresses small shrinking, excellent crack resistance, because of dense surface structure and little water absorption. It also shows excellent workability of blended concrete.

4.6.5 Composite Portland Cement (P.C)

Composite Portland cement refers to the Portland cement which is produced by Portland cement clinker and two or more kinds of mixing materials. Composite Portland cement has 32.5-52.5MPa strength grades. The requirements of strength are shown in Table 4.5. The other properties of composite Portland cement are the same with P.O.

Table 4.5 Strength demands of P.C

Strength grade	Compressive strength/MPa		Bending strength/MPa	
	3 d	28 d	3 d	28 d
32.5	11.0	32.5	2.5	5.5
32.5R	16.0	32.5	3.5	5.5
42.5	16.0	42.5	3.5	6.5
42.5R	21.0	42.5	4.0	6.5
52.5	22.0	52.5	4.0	7.0
52.5R	26.0	52.5	5.0	7.0

4.6.6 Other Types of Cement

1. Expansive Cement

Although not new, the first commercial use of expansive cement in the United States was on the West Coast in 1963. The purpose of the cement is to compensate for normal drying shrinkage. There are three types of expansive cements:

(1) Type 1. A mixture of Portland cement plus anhydrous calcium sulphoaluminate ($4CaO \cdot 3AlSO_3 \cdot SO_3$), calcium sulfate ($CaSO_4$), and lime ($CaO$).

(2) Type 2. A mixture of Portland cement, calcium aluminate cement, and calcium sulfate.

The compounds of type 1 and type 2 expansive cements may be burned separately and the clinker ground with the Portland cement clinker or added to the kiln and burned with the Portland cement clinker.

(3) Type 3. A Portland cement containing excess amount of tricalcium aluminate and calcium sulfate.

2. White Cement

White cement is made by carefully selecting raw ingredients to eliminate or greatly reduce iron and manganese, which gives Portland cement its gray color. White cement has essentially the same concrete-making properties as regular Portland cements.

3. Aluminous Cement

Aluminous cements are made by mixing bauxite rather than siliceous materials with limestone in the kiln. Calcium aluminates are composed of mainly $CaO \cdot Al_2O_3$ rather than calcium silicates. This type of cement is useful for refractory mortar, for rapid set and high early strength, and in areas where chemical attack is a problem.

A mixture of aluminous cement and Portland cement will give a rapid set in concrete. The time of set can be regulated by varying the proportions of Portland and aluminous cements.

4. Low-Alkali Cement

The term "low-alkali cement" refers to the amount of alkalis, sodium and potassium oxide (Na_2O and K_2O), found in Portland cement. Calcium oxide, one of the principal compounds of Portland cement, reacts with SiO_2 to produce hardened cement. Sodium and potassium oxides also combine with amorphous SiO_2, but the silica gel produced is unstable in concrete and causes expansion and cracking.

The term "alkali-silica reaction" refers to the reaction between Na_2O and K_2O and aggregate particles containing amorphous SiO_2.

Specifications for low-alkali cement limit $Na_2O + 0.658\ K_2O$ to a certain low value as a principal precaution against alkali-aggregate reaction.

Questions

4.1 What ingredients are used for the production of Portland cement?

4.2 What are the main chemical components in cement? What influence do minor compounds bring to cement?

4.3 Describe the steps of setting and hardening of cement and the reasons of false setting.

4.4 What is soundness of cement? What kind of factors can cause the destructive expansion?

4.5 Briefly introduces several common cement types and their properties.

References

[1] A J FRANCIS, DAVID, CHARLES. The Cement Industry, 1796–1914: A History[J]. Business History Review, 1979, 53(3):430-431.

[2] B FAGERLUND GILLBERG, G JÖNSSON, Å TILLMAN, et al. Betong och miljö [Concrete and environment] (in Swedish) [M]. Stockholm: AB Svensk Byggtjenst, 1999.

[3] D L RAYMENT. The electron microprobe analysis of the C-S-H phases in a 136-year-old cement paste[J]. Cement and Concrete Research, 1986.

[4] HENRY REID. A practical treatise on the manufacture of Portland cement [M]. London: E & F.N. Spon, 1868.

[5] HENRY REID. The Science and art of the manufacture of Portland cement with observations on some of its constructive applications[M]. London: E&F.N. Spon, 1877.

[6] MEADE, RICHARD KIDDER. Portland cement: its composition, raw materials, manufacture, testing and analysis [M]. Easton: The Chemical Publishing, 1926.

[7] ROBERT COURLAND. Concrete planet: the strange and fascinating story of the world's most common man-made material [M]. Amherst, N.Y.: Prometheus Books, 2011.

[8] ROBERT G BLEZARD. The History of Calcareous Cements[J]. Leas Chemistry of Cement & Concrete, 1998:1-23.

[9] SAIKIA MIMI DAS, BHARGAB MOHAN DAS. Elements of Civil Engineering [M]. New Delhie: PHI Learning Private Limited, 2010.

[10] THOMAS F HAHN, EMORY LELAND KEMP. Cement mills along the Potomac River [M]. Morgantown, WV: West Virginia University Press, 1994.

Chapter 5

Portland Cement Concrete

5.1 Introduction

Concrete is a stone-like material obtained by permitting a carefully proportioned mixture of cement sand and gravel or other aggregate and water to harden in forms of the shape and dimensions of the desired structure.

Concrete is widely used for making architectural structures, foundations, brick/block walls, pavements, bridges/overpasses, highways, runways, parking structures, dams, pools/reservoirs, pipes, footings for gates, fences and poles, and even boats. Concrete is used in large quantities almost everywhere mankind has a need for infrastructure.

The amount of concrete used worldwide, ton for ton, is twice that of steel, wood, plastics, and aluminum combined. Concrete's use in the modern world is exceeded only by that of naturally occurring water. Given the size of the concrete industry, and the fundamental way concrete is used to shape the infrastructure of the modern world, it is difficult to overstate the role this material plays today.

5.1.1 Characteristics of Concrete

(1) Convenient for use: the new mixtures have good plasticity that can be cast into components and structures in various shapes and sizes.

(2) Cheap: raw materials are abundant and available. More than 80% of them are sand and stone whose resources are rich, energy consumption is low, according with the economic principle.

(3) High-strength and durable: the strength of ordinary concrete is 20-55 MPa with good durability.

(4) Easy to be adjusted: the concrete with different functions can be made just by changing the varieties and quantities of composing materials to meet various demands of projects; steel bar can be added to concrete to improve its strength, and this kind of concrete is a composite material (reinforced concrete) which can improve its low tensile and bending strength in order to meet the needs of various structural engineering.

(5) Environment-friendly: concrete can make full use of industrial wastes, such as slag,

fly ash and others to reduce environmental pollution.

Its major shortcomings are high dead weight, low tensile strength, brittle and easy to crack.

5.1.2 Classification of Concrete

1. In Portland Cement Concrete

The binder is a mixture of Portland cement and water. Asphalt and other cements are used to make various types of concrete, but commonly the term "concrete" refers to Portland cement concrete.

2. By Apparent Density

Concrete includes three types:

Heavy concrete: ρ_0>2600 as shielding materials of atomic energy engineering.

Normal concrete: ρ_0=2000-2500 in several bearing structure.

Lightweight concrete: ρ_0<1900 including light aggregate concrete and porous concrete.

3. By Strength

Concrete includes three types: ordinary concrete, high strength concrete, super strength concrete.

Ordinary concrete: compressive strength <60 MPa.

High-strength concrete: compressive strength>60 MPa.

It is applied largely to the high-rise building, large span bridges and high-strength prefabrication components and so on.

Super-strength concrete: compressive strength>80 MPa.

4. By Forming or Construction Technology

It includes four kinds of concretes: deposit concrete, precast concrete, premixed concrete and shotcrete concrete.

5.2 Composition of Concrete

Concrete is made up of paste (cement water), aggregate (sand, gravel) and admixture. Generally, the amount of sand and stone accounts for above 80% of the total volume, functioning as frame, so they are respectively called as fine aggregate and coarse aggregate. Mixed with water, cement becomes cement paste, and cement mortar not only wraps the surface of particles and fills their gaps, but also wraps stones and fills their gaps, then concrete coming into being (Figure 5.1). Cement paste can function as greasing before hardening, which renders concrete mixture with good mobility; after hardening, aggregates stick together and form a hard entity, known as man-made stone-concrete.

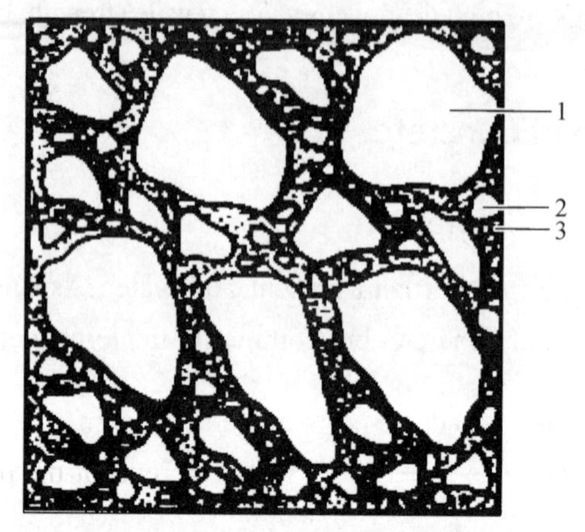

1-coarse aggregate; 2-fine aggregate; 3-cement paste.

Figure 5.1 The cement structure

5.2.1 Cement

Cement is the most important component for concrete and relatively expensive. There are many types of concrete, based on different cements. In the preparation of concrete, the choice of cement varieties and strength grades are directly related with the durability and economy of concrete.

1. Selection of Types

Select according to the different environment. The six general cements are commonly used in Portland cement, ordinary Portland cement, furnace-slag cement, Portland pozzolan cement, Portland fly-ash cement and composite Portland cement. The selection principles for the six common kinds of cement can be referred to in chapter four.

2. Selection of Grade

The cement strength grades are corresponding to the design strength grades of concrete. For ordinary concrete the cement is 1.5-2 times the strength of the concrete. For high strength concrete, it is 0.9-1.5 times the strength of the concrete.

5.2.2 Aggregate

In cement concrete, 60%-75% of the volume and 79%-85% of the weight are made up of aggregates. The aggregates act as a filler to reduce the amount of cement paste needed in the mix. In addition, aggregates have greater volume stability than the cement paste. Therefore, maximizing the amount of aggregate, to a certain extent, improves the quality and economy of the mix.

Generally, the aggregates used for ordinary concrete can be divided into two types by their sizes. The one whose diameter is more than 5.00 mm is called coarse aggregate, and the one whose diameter is less than 5.00 mm is called fine aggregate.

1. Fine Aggregate

The fine aggregates used in ordinary concrete generally are the natural sand which comes into being when the natural rock (excluding soft rock and weathered rock) has experienced natural weathering, water transportation, sorting, stacking, and other kinds of natural conditions; the machine-made sand (is made by grinding and sorting by machine, and the diameter of the rock particles is less than 5.00 mm, except the particles of soft rock and weathered rock.) through de-dust treatment and the mixed sand (made by mixing machine-made sand and natural sand) are collectively call manufactured sand.

According to different sources, natural sand can be divided into river sand, sea sand, mountain sand and desalted sea sand.

In national standards *Sand for Building* (GB/T 14684-2001) and *Pebble and Crushed Stone for Building* (GB/T 14685-2001), it is regulated that the sand for building can be classified into category I, category II , and category III based on the technical quality requirements.

Category I: those used in the concrete whose strength grade is more than C60;

Category II: those used in the concrete whose strength grade is between C30-C60;

Category III: those used in the concrete whose strength grade is less than C30.

(1) Grain Gradation and Coarseness of Sand

The grain gradation and coarseness of sand are determined by screen residue analysis. Grading region and fineness modulus can be used to express the gradation and the coarseness of sand particles respectively. The relationship between the cumulative screen residue (which refers to the screen residue of one sieve to the sum of all the unit screen residue percentages whose sieves are thicker than it) and the unit screen residue (which means the mass of screen residue to the mass of the total sample sand) is shown in Table 5.1.

Table 5.1 The relationship between cumulative screen residue and unit screen residue

Sieve hole/mm	Unit screen residue/%	Cumulative screen residue/%
4.75	a_1	$A_1=a_1$
2.36	a_2	$A_2=a_1+a_2$
1.18	a_3	$A_3=a_1+a_2+a_3$
0.60	a_4	$A_4=a_1+a_2+a_3+a_4$
0.30	a_5	$A_5=a_1+a_2+a_3+a_4+a_5$
0.15	a_6	$A_6=a_1+a_2+a_3+a_4+a_5+a_6$

According to GB/T14684-2001, there are three grading regions of sand when it is calculated by the percentage of the cumulative screen residue of 0.63 mm hole sieve, shown in Table 5.2.

Table 5.2 Grading regions of sand particles (GB/T 14684-2001)

Hole Size/mm	Grading Regions		
	I	Ⅱ	Ⅲ
	Cumulative Screen Residue/%		
10	0	0	0
5	10-0	10-0	10-0
2.5	35-5	25-0	15-0
1.25	65-35	50-10	25-0
0.63	85-71	70-41	40-16
0.315	95-80	92-70	85-55
0.16	100-90	100-90	100-90
0.16	100-90	100-90	100-90

Note: ① Three grading regions are compartmentalized by accumulative total sieve residue percent of sieve size with 0.63 mm.

② Compared with the numbers listed in the table, the accumulative total sieve residue percent of the sand can go beyond 5% of all the sand except 5 mm and 0.63 mm.

Grading region grading curve of sand is shown in Figure 5.2.

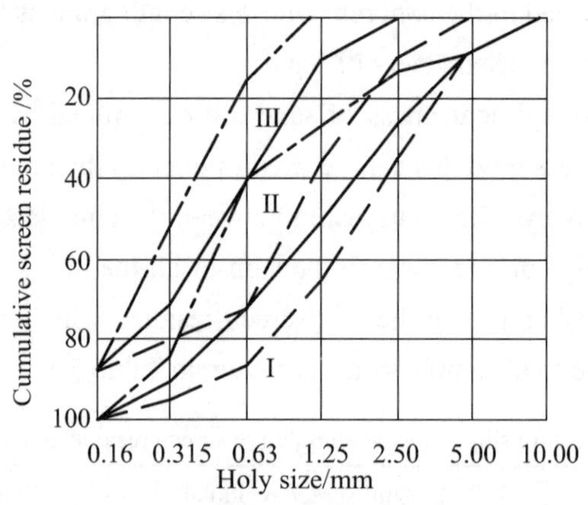

Figure 5.2 Grading region grading curve of sand

The coarseness of sand is expressed by fineness modulus (μ_f), defined as follows:

$$\mu_f = \frac{(A_2 + A_3 + A_4 + A_5 + A_6) - 5A_1}{100 - A_1}$$

The bigger the fineness modulus is, the coarser the sand is. According to fineness modulus, we can decide the types of fineness referring to Table 5.3.

Table 5.3 Types of fineness

Types of fineness	μ_f
coarse sand	3.1-3.7
medium sand	2.3-3.0
fine sand	1.6-2.2
super fine sand	0.7-1.5

It should be reminded that the fineness modulus cannot reflect the quality of their grading regions. The sand with the same fineness modulus can have very different grading regions. Therefore, the particle gradation and the fineness modulus should be considered in the preparation of concrete.

(2) Requirements of Aggregate

① Impurities

Impurities hinder the hydration or causes harden cement paste corrosion, and reduce the bond between cement paste and aggregate.

Types: mica, clay, silt and organic substance.

Damage: hinder the bond between paste and aggregate, weaken the strength of concrete, increase the requirement quantity of water, increase the shrinkage of concrete and bring corrosion to harden cement paste.

Processing methods: wash impurities out as requested, if too much in the sand.

The amount of impurities must be in agreement with GB/T 14684-2001(JGJ 52-92) (Table 5.4).

Table 5.4 Amounts of impurities (GB/T 14684-2001)

Type	Index		
	I	II	III
Clay content/%	<1.0	<2.0	<5.0
Clod content/%	0	<1.0	<2.0
Mica/%	<1.0	<2.0	<2.0
Light matter/%	<1.0	<1.0	<1.0
Organic substance (colorimetry)	Eligible	Eligible	Eligible
Sulfides and sulfates/%	<0.5	<0.5	<0.5
Chloride/%	<0.01	<0.02	<0.06

② Particle Shape and Surface Texture

Particle shape, surface texture and their influences in properties are listed in Table 5.5.

Table 5.5 Particle shape and surface texture

Type	Particle Shape	Surface Texture	Workability	Strength
River sand	Round or elliptical	Lubricity	Well	Low
Sea sand				
Mountain sand	Pointed	Rough	Bad	High

2. Coarse Aggregate

According to shape of primary coarse aggregate, it can be divided into gravel and pebble. Gravel is the widely used primary coarse aggregate.

By quality, there are three types: Ⅰ, Ⅱ and Ⅲ. Ⅰ: those used in the concrete whose strength grade is more than C60; Ⅱ: those used in the concrete whose strength grade is between C30-C60; Ⅲ: those used in the concrete whose strength grade is less than C30.

(1) The Grain Composition of Stone and the Maximum Particle Diameter

The grain composition of stone includes continuous size fraction and single size fraction which are also determined by sieve analysis. The determination method is the same with that of fine aggregate.

Continuous size fraction has priority in the design of concrete mixture ratio. Single size fraction can be used in the composition of required continuous size fraction and also used with continuous size fraction to improve the grading or to prepare the continuous size fraction of larger-sized particles. The single size fraction should not be used "singly" in the preparation of concrete. If it must be used lonely, the technical and economic analysis should be made and the experiment should be conducted to prove that there will be no segregation or any impact on the quality of concrete.

The maximum nominal size of stone particle is the maximum particle diameter of this size fraction. If the maximum diameter increases, its total area decreases when quality remains the same. Thus, from the economic perspective, cement can be saved by increasing the maximum diameter. Therefore, the bigger maximum size should be chosen if the circumstance allows.

(2) Requirements of Aggregate

① Impurities

Clay content, clod content and the content of needle shape particles and slice shape particles (Figure 5.3) must be in agreement with GB/T 14685-2001 (JGJ 52-92).

Figure 5.3 Needle and slice shape particles

Dctails of alkali-aggregate reaction will be introduced later in durability of concrete. If there is potential danger, the alkali content should be less than 0.6%. A special test is needed when we use the admixture with potassium and sodium.

② Particle Shape and Surface Texture

The particle shape and surface texture is the same with that of the sand.

③ Aggregate Strength

There are two methods to measure the strength of the aggregate strength: cubic strength of rock and crushing index of gravel.

The compressive strength of the cube is measured in saturated water. For ordinary concrete: Aggregate strength $\not< 1.5 f_{cu}$ for HSC: Aggregate strength $\not< 2.0 f_{cu}$.

Also select the strength of rock according to the rocks types (Table 5.6).

Table 5.6 Aggregate strength of different types

Rocks	Strength
Igneous rock	$\not<$80 MPa
Metamorphic rock	$\not<$60 MPa
Aqueous rock	$\not<$30 MPa

(3) Crushing Index of Gravel

Method: put the gravel with the diameter of 10-20 mm into the standard cylinder three times, press them to 200 kN and shift them in 2.5 mm sieve.

$$Crushing\ index = \frac{m_0 - m_1}{m_0}$$

In this formula: m_0 is the mass of dry gravel before crushing;

m_1 is the mass of dry gravel after crushing and sifting.

Crushing index of gravel or scree is listed in Table 5.7 and Table 5.8.

Table 5.7 Crushing index of gravel of different types

Types of rock	Concrete strength grade	Gravel crushing index/%
Aqueous rock	C60-C40	≤10
	≤C35	≤16
Metamorphic rock or plutonic igneous rock	C60-C40	≤12
	≤C35	≤20
Igneous rock	C60-C40	≤13
	≤C35	≤30

Table 5.8 Crushing index of scree (GB/T 14685-2001)

Concrete strength grade	C60-C40	≤C35
Crushing index/%	≤12	≤16

5.2.3 Water for Concrete

Combining water with a cementitious material forms a cement paste by the process of hydration. The cement paste glues the aggregate together, fills voids within it, and makes it flow more freely. However, impure water used to make concrete can cause problems when setting or in causing premature failure of the structure. We should use drinking water and clean natural water for mixing and conserving concrete.

The substance content in water for concrete should accord with the limits of JGJ 63-89 in Table 5.9.

Table 5.9 Limits of the substance content in water for concrete

Types	Prestressed concrete	Reinforced concrete	Plain concrete
pH	>4	>4	>4
Insoluble matter/(mg/L)	<2000	<2000	<5000
Soluble matter/(mg/L)	<2000	<5000	<10000
Cl^-/(mg/L)	<500	<1200	<3500
SO_4^{2-}/(mg/L)	<600	<2700	<2700
S^{2-}/(mg/L)	<100	-	-

5.2.4 Concrete Admixture

Chemical admixtures are materials in the form of powder or fluids that are added to the concrete to give it certain characteristics not obtainable with plain concrete mixes. In normal use, admixture dosages are less than 5% by mass of cement and are added to the concrete at the time of batching/mixing. The common types of admixtures are as follows.

1. Water-reducing Agent

Water-reducing agent refers to the admixture used for reducing water consumption and strengthening functions when the slump degrees of mixtures are basically the same. By raw materials and chemical components, water-reducing admixtures can be divided into: lignin sulfonate, alkyl aryl sulfonates (commonly known as coal tar water-reducing admixture), sulfonated melamine-formaldehyde resin sulfonate (commonly known as melamine water-reducer), molasses and humic aid water-reducer, and others. Based on performances and functions, water-reducing admixtures can be divided into: ordinary water-reducer, effective water-reducer, hardening water-reducer, retarder water-reducer, and air entraining water-reducer.

2. Air Entraining Admixture

Air entraining admixture refers to the admixture that entrains a large number of uniform, stable and closed tiny bubbles in the process of mixing concrete to reduce the segregation of concrete mixture, improve the workability, and also enhance anti-freeze ability and durability of concrete. It is a kind of surfactant, too. It has influences on concrete as follows:

(1) It can improve the workability of concrete mixtures.

(2) It can enhance impermeability and frost resistance.

(3) It can reduce strength. If the air content in concrete increases by 1%, its compressive strength will decrease by 4%-6%. Thus, the adding amount of air entraining admixture should be appropriate.

3. Hardening Accelerator

Hardening accelerator refers to the admixture that can accelerate the development of early strength of concrete. It applies to the construction that is constructed in winter, emergency engineering and time-limited ones. The use of hardening accelerator can make C20 reach demoulding strength within 16 hours and the strength allowing floor slab installment on it within 36 hours so as to speed up the construction.

4. Set Retarder

Set retarder refers to the admixture that can delay the setting time of concrete mixing materials, and have no bad impact on the development of concrete's latter strength. It often contains lignin sulfonate, carbohydrate, inorganic salts and organic acids. The most common ones are calcium lignin sulfonate and molasses. And the retardant effect of molasses is better. Set retarder is appropriately used in the projects that need to delay time, such as high temperature or long transport distance, to prevent the loss caused by the early slump of concrete mixtures; and also for the layer pouring concrete, set retarder is often added to prevent cold joint. In addition, set retarder can be added into mass concrete to extend the heat-releasing time.

5. Anti-freeze

Anti-freezing admixture makes concrete gain strength normally in negative temperature and plays the role of freezing point, anti-freezing and accelerating strength gain in concrete. The admixture used in the construction is composite anti-freezing admixture which is compounded of anti-freezing component, early strength component, reducing water component, even air entraining component so as to improve the effect on anti-freezing.

6. Rust-resistant Agent

Rust-resistant agent is the admixture that can retard the corrosion to steel bars in concrete or other embedded metal, also called corrosion inhibitor. The common agent is sodium nitrite. Some admixtures contain chloride salt which will corrode steel bars (thus, it is necessary to control the content of chloride ions), so the adding of rust-resistant agent can retard corrosion to steel bars for the sake of protection.

5.3 The Main Technical Properties of Concrete

5.3.1 Workability of Fresh Concrete

Fresh concrete refers to the mixtures made by cement, sand, stone and water in a certain proportion that has not yet hardened.

1. The Concept of Workability

Workability is the ability of a fresh (plastic) concrete mix to fill the form/mold properly with the desired work (vibration) and without reducing the concrete's quality. The workability of concrete mixture is an integrated technical property, including mobility, viscidity, and water retention, the three aspects. Mobility is the property that fresh concrete can move and fill every corner of the mould evenly and densely under the role of deadweight and mechanical vibration. Viscidity refers to the property that there exists a certain bonding power among the components of concrete mixture to avoid stratification and segregation and maintain the entire uniformity. Water retention is the property to prevent moisture from precipitating during the construction process of fresh concrete.

2. Evaluation of Workability

Mobility has great impact on the properties of mixture through actual analysis. Thus, mobility is often used to test concrete mixture assisted by observation of viscidity and water retention to judge whether fresh concrete can satisfy the demand of projects.

(1) The Determination of Slump

Workability can be measured by the concrete slump test, a simplistic measure of the plasticity of a fresh batch of concrete following the ASTM C143 or EN 12350-2 test standards. Slump is normally measured by filling an "Abrams cone"(Figure 5.4) with a sample from a fresh batch of concrete. The cone is placed with the wide end down onto a level, non-absorptive surface. It is then filled in three layers of equal volume, with each layer being tamped with a steel rod to consolidate the layer. When the cone is carefully lifted off, the enclosed material slumps a certain amount, owing to gravity. A relatively dry sample slumps very little, having a slump value of one or two inches (25 or 50 mm) out of one foot (305 mm). A relatively wet concrete sample may slump as much as eight inches.

(2) The Determination of Vebe Consistometer

The mobility of the concrete mixture whose slump is less than 10 mm should be denoted by Vebe consistometer.

The determination method is: put the composite of truncated cone shape in the container of Vebe consistometer instrument and make the transparent disc touch to the top of cone, then put on vibrating table and record the time when the disc is covered with paste.

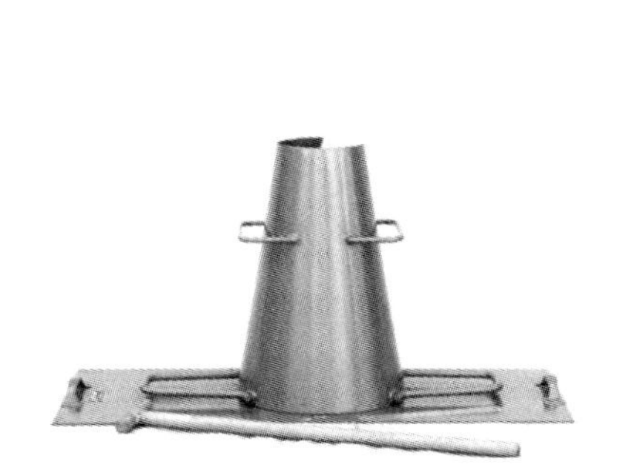

Figure 5.4 Slump cone

Figure 5.5 Vebe consistometer instrument

3. The Major Factors Influencing Workability

(1) The Quantity of Cement and Water-cement Ratio

The mobility of concrete mixture depends on cement paste. In unit volume, the more the cement paste, the greater the mobility, if the water-cement ratio is the same. However, if cement paste is too much, pulp fluid will occur; and if cement paste is too little, the cohesion among aggregate particles will be small, which may lead to segregation and collapse. If the using amounts of cement and aggregate are the same, increasing the water-cement ratio can strengthen the mobility; on the contrary, the mobility will be lowered. Still, if the water-cement ratio is too big, the cohesion and water retention of the mixture will decrease; if the ratio is too small, the mobility will decrease, which may impact projects. Thus, the ratio should be chosen reasonably according to the strength and durability of concrete. It should be noted that both the impact of cement paste and that of water-cement ration are basically the impact of water consumption. Therefore, the key factor to affect the workability of concrete mixture is the water consumption per unit volume.

(2) Sand Percentage

Sand percentage refers to how large the quality of sand within concrete is relative to the total mass of sand and stone. Sand percentage has great influence on mixture. It determines the void and surface area of aggregates (W and C are fixed). With low S_p, it lacks of mortar so it results in poor lubrication of the surface of aggregates and poor mobility, which tends to segregate. With high S_p in large surface area, cement mortar wraps up sand and fills paste, which results in low mobility. Optimized S_p: when W and C are fixed, the optimized S_p makes the composite gain the highest mobility with favorable viscidity and water retention, or makes the composite gain the required mobility with the least cement used.

(3) Properties of Components

It is the categories and fineness of cement that mainly influence the workability of mixture. If Portland cement and ordinary cement are used, the mobility will be big and the water retention will be good; if slag cement and pozzolana cement are used, the mobility will be small and water retention will be bad; the mobility of fly-ash cement is better than that of ordinary cement, and the water retention and cohesion are good. In addition, the fineness of cement also has influence on the workability of mixture. The finer the cement is ground, the smaller the mobility is, but the cohesion and water retention will be good.

Aggregate can affect the workability of mixtures, mainly including: gradation, particle shape, surface characteristic, and particle diameter. Generally, the aggregate with good gradation has big mobility and good cohesion and water retention; the aggregate (river sand, gravel, etc.) whose surface is smooth has big mobility; the increase of the diameter and decrease of the total surface area will increase the mobility.

(4) Time and Temperature

By hydration and vaporization, water is absorbed by aggregates and mobility decreases accordingly.

The relationship between slump and time is shown in Figure 5.6. Slump decreases while time increases. Slump test should be preceded in 15 minutes after preparation of the composite in construction.

Mobility decreases with temperature rising. Temperature must be noticed to get a required workability. More mixed water in summer should be used than in winter.

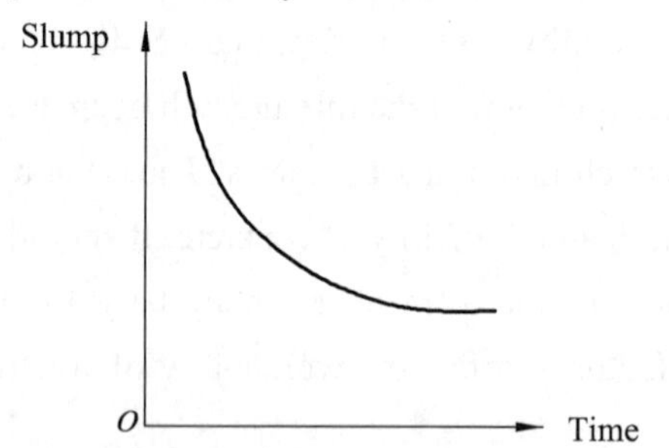

Figure 5.6 Relationship between slump and time

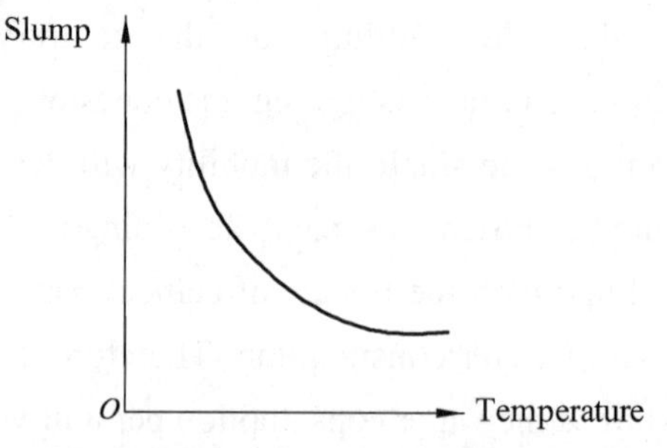

Figure 5.7 Relationship between slump and temperature

(5) Test Condition

Test conditions also have effects on the mobility of the concrete composite, such as device condition-parameter, concrete condition-humidity, smoothness, etc.

4. Approach to Improve Workability

(1) When slump is too small, improve the quantity of cement paste for the mobility with

the stability of W/C.

(2) When slump is too large, improve the quantity of aggregates to decrease the mobility with the stability of S_p.

(3) Choose the optimized S_p.

(4) Improve the gradation of aggregates.

(5) Choose coarse aggregates if possible.

(6) Use admixture.

5.3.2 The Strength of Hardened Concrete

1. Compressive Strength and Intensity Grade of Concrete

The strength of concrete includes compressive strength, tensile strength, bending strength, and shear strength, among which the compressive strength is the biggest one. Thus, concrete is mainly used to bear pressure.

Based on *Test Method of Mechanical Properties on Ordinary Concrete* (GBJ 81), the national standard, the compressive strength of concrete cube (simply called compressive strength of concrete) is the compressive strength value obtained as follows: conserve the cube specimen with side length of 150 mm made through the standard method under standard conservation conditions [temperature of (20 ± 3)°C, and relative humidity above 90% or in water], and test and calculate it by standard method to get the compressive strength value which is called the compressive strength of concrete cube (expressed by f_{cu}). For the cube specimens with non-standard sizes (with the side length of 100 mm or 200 mm), the result should be multiplied by conversion coefficient and be converted into the strength value of standard specimen. Thus the cube specimen with side length of 100 mm should be multiplied by 0.95; and that with 200 mm length should be multiplied by 1.05.

In order to facilitate the design selection and construction, concrete is divided into several grades, namely strength grades. They are divided in accordance with the compressive strength values of cubes ($f_{cu,k}$). Ordinary concrete is usually divided into 9 grades: C15, C20, C25, C30, C35, C40, C45, C50, and C55 (the concrete whose strength grade is≥C60 is called high-strength cement).

2. Other Strength

(1) Prism Compressive Strength f_{cp}

The compressive strength of 150 mm×150 mm×300 mm prism is cured in standard condition in age of 28 days.

The relations of Prism compressive strength and compressive strength are: f_{cp} =0.7-0.8f_{cu}.

(2) Tensile Strength

Tensile strength is tested through split tensile test.

(3) Principle

Figure 5.8 is the ketch map of split tensile test. Apply linear load distributed evenly on two opposite surfaces, thus the tension stress will be produced on the vertical surface affected by the external force. This force will be calculated by the elastic theory.

(4) Formula

Split tensile test strength f_{ts}:

$$f_{ts} = \frac{2P}{\pi A} = 0.637\frac{P}{A}$$

In this formula: P is the damage load (N);

A is the split area of specimen (mm²).

(5) Bending Strength (Figure 5.9)

Middle third point loading experiment test;

Specimen: 150 mm×150 mm×600(550) mm beam specimen.

$$f_m = \frac{PL}{bh^2}$$

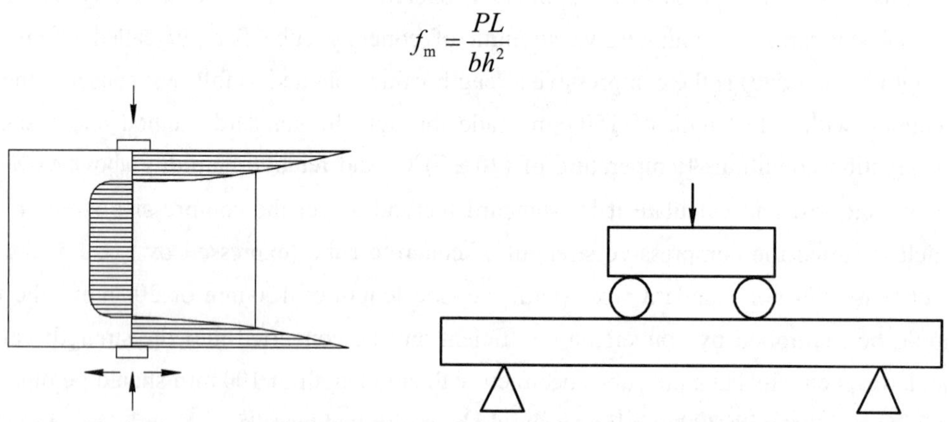

Figure 5.8 The ketch map of split tensile test   Figure 5.9 Sketch map of bending test

3. The Major Factors Influencing the Compressive Strength

There many factors influencing the compressive strength of concrete, such as the strength of cement, the bond strength between aggregate and cement, the quality of materials, the mixing ratio of materials, and the construction conditions. But the major factors are listed in the following:

(1) The Influence of Cement Strength Grades and Water-cement Ratio

Cement strength grades and water-cement ratio are the main factors impacting concrete strength. The chemically combined water needed in cement hydration generally account for 23% of the mass of cement. But in the actual mixture of concrete, more water is needed to obtain greater mobility. The space occupied by excessive water will turn into pores after hardening which will lower the density and strength of concrete.

It is proved that the smaller the water-cement ratio is, the higher the strength of cement will be, the higher the cohesive power will be, and the strength of concrete will be, under the same condition.

A large number of tests have proved that: at the age of 28 d, the relationship between the concrete strength ($f_{cu,0}$), the actual strength of cement (f_{ce}) and the water-cement ratio (W/C) is in line with the following formula:

$$f_{cu,0} = \alpha_a f_{ce}\left(\frac{C}{W} - \alpha_b\right)$$

In the formula: α_a and α_b are the regression coefficients. If crushed stones are used, α_a =0.46 and α_b =0.07; if gravels are used, α_a =0.48 and α_b =0.33.

(2) The Influence of Aggregate

The strength of coarse aggregate, particle size and gradation are important to the concrete strength.

① When the aggregate strength is high, the crack expands and passes over the interface between aggregate and cement, thus the concrete strength is increased.

② surface texture: W/C>0.65, no effect; $W/C < 0.4$, f_{cu}=1.38$f_{cu,k}$.

③ DMAX has smaller effect on ordinary concrete. When HSC and DMAX goes up, strength will drop (effect of dimension).

Obviously aggregates play an important role in producing strength in concrete. The shearing strength of an aggregate particle itself may control the strength across a shear plan when the bond between the aggregate and the paste is strong enough to force the shear plane through the aggregate particles rather than around them.

(3) The Relationship Between Age and Strength

Under normal curing conditions, the strength of concrete increases with its age. During the initial 3-7 d, it grows fast and it can reach the numerical value of the regulated design strength. Afterwards, it grows gradually slowly, even unchangeable ever after. The growth of concrete strength is in direct proportion to the logarithm of age (which is more than 3 d) under standard curing conditions, calculated as follows:

$$f_n = f_{28}\frac{\ln n}{\ln 28}$$

In the formula: f_{28} is the concrete compressive strength of 28 d;

f_n is the concrete compressive strength of nd ($n \geqslant 3$).

The above formula applies to the concrete made of cement in media grades under standard curing conditions. The real situation, however, is complicated, so it is generally only taken as reference.

(4) The Influence of Curing Temperature and Humidity

The concrete strength is influenced by the degree and speed of cement hydration, which is affected by the humidity and temperature. Higher the temperature is, faster the speed of cement hydration is, higher the concrete strength is. The larger the humidity, the higher degree of cement hydration is. Relationship among strength, temperature and damping age of concrete are shown in Figure 5.10 and Figure 5.11.

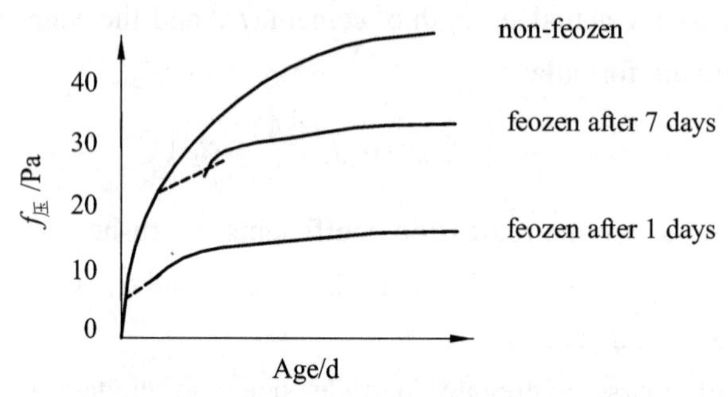

Figure 5.10　Influence of temperature to strength

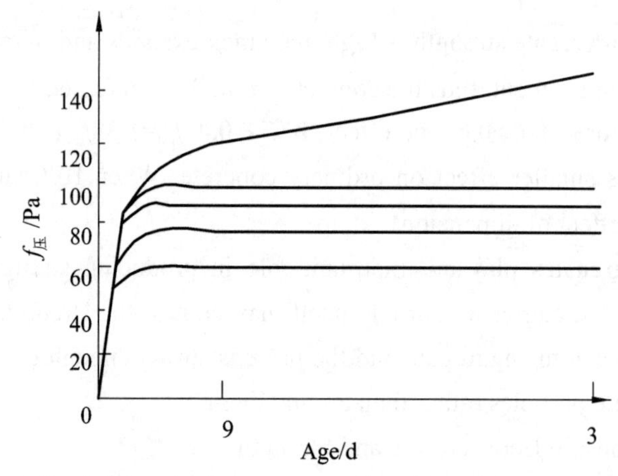

Figure 5.11　Relationship between strength and damping age of concrete

The fixed temperature and relative humility in a certain time will guarantee the normal development of strength and other performances of concrete after the formation of concrete. There are three types of curing: natural cure, steam cure and autoclaved cure.

(5) Influence of Admixtures on Strength

Admixtures influence the strength of concrete in different ways. Air-entraining agents reduce strength; water-reducing admixtures increase strength by permitting a reduction in the water-cement ratio. Accelerating agents produce higher early strength, and retarding agents delay the initial strength gain.

4. Approaches to Improve Strength

(1) High-grade cement and quick hardening and high early strength cement.

(2) Using dry and hard concrete makes porosity decrease and density improved.

(3) Hydrothermal treatment can improve efficiency, save space and increase strength.

(4) Employing the mechanic mixing, vibration, strong mixing and high frequency vibration.

(5) Blending agent and mineral admixture; predicting the concrete development; producing HSC, HPC and so on.

5.3.3 Early Volume Change

When the cement paste is still plastic it undergoes a slight decrease in volume of about 1%. This shrinkage is known as plastic shrinkage and is due to the loss of water from the cement paste, either from evaporation or from suction by dry concrete below the fresh concrete. Plastic shrinkage may cause cracking (Figure 5.12); it can be prevented or reduced by controlling water loss.

In addition to the possible decrease in volume when the concrete is still plastic, another form of volume change may occur after setting, especially at early ages. If concrete is not properly cured and is allowed to dry, it will shrink. This shrinkage is referred to as drying shrinkage, and it also causes cracks. Shrinkage takes place over a long period of time, although the rate of shrinkage is high early, then decreases rapidly with time. In fact, about 15% to 30% of the shrinkage occurs in the first two weeks, while 65% to 85% occurs in the first year. Shrinkage and shrinkage-induced cracking are increased by several factors, including lack of curing, high water-cement ratio, high cement content, low coarse aggregate content, existence of steel reinforcement, and aging. On the other hand, if concrete is cured continuously in water after setting, concrete will swell very slightly due to the absorption of water. Since swelling, if it happens, is very small, it does not cause significant problems. Swelling is accompanied by a slight increase in weight.

The degree of shrinkage depends on the size and shape of the concrete structure. Also, nonuniform shrinkage could happen due to the nonuniform loss of water. This may happen in mass concrete structures, where more water is lost at the surface than at the interior. In cases such as this, cracks may develop at the surface. In other cases, curling might develop due to the nonuniform curing throughout the structure and, consequently, nonuniform shrinkage.

Figure 5.12 Plastic shrinkage cracking

5.3.4 Durability of Concrete

Durability refers to that concrete can resist the influence of exterior corrosive substance and maintain good usability and complete appearance so that it can maintain the safety and usability of the structure. That is to say that concrete can maintain stable quality after being used for a long time. Here are some common issues concerning durability.

1. Frost Resistance of Concrete

The frost resistance of concrete is the property that concrete resist the role of freeze-thaw cycles without damage when it suffers frost in the saturated state.

Frost resistance is indicated by frost-resistance level. It is determined by the maximum cycle number of the 28d-age specimen whose mass loss and the strength loss do not exceed 5% and 25% respectively, when it stays in repeated freeze-thaw cycle of −15-−20 °C and 15-20 °C in the saturated state. The frost-resistance levels can be divided into F50, F100, F150 and above F150, which respectively indicate that the cycle numbers borne by concrete are not less than 50, 100, 150 and above.

The main factors affecting the frost resistance of concrete are varieties of cement, water-cement ratio and the sturdiness of aggregates. The measures to improve frost resistance are to enhance density, reduce water-cement ratio and mix air-entraining agent or air-entraining and water-reducing admixture.

2. Impermeability of Concrete

The impermeability of concrete refers to the ability to resist the permeability from compressive water. It is one of the most important symbols of durability. It influences the anti-freezing and anti-corrosion of the concrete directly.

The key to improve the impermeability of concrete is to enhance density and improve the inner pores structures of concrete. There are specific measures to reduce the water-cement ratio, adopt water-reducing agent, add air-entraining agent, select dense, clean and good aggregates, and strengthen curing.

3. Anti-corrosion of Concrete

If the water surrounding concrete is corrosive, corrosion must be attached importance to. Environmental erosion mainly refers to the corrosion of cement paste. The chloride ions in sea water can also corrode steel bars to accelerate the damage of concrete. The anti-corrosion of concrete is improved by selecting appropriate cement varieties and enhancing the density of concrete.

4. Carbonization of Concrete

The carbonization of concrete is the process that carbon dioxide in the air penetrates concrete, chemically reacts with calcium hydroxide in cement paste and generates calcium carbonate and water to reduce the alkalinity of concrete, also known as neutralization. It is known that cement generates a lot of calcium hydroxide in the process of hydration, which fills the pores in concrete with saturated calcium hydroxide solution, pH value of 12-13. In such alkaline environment, there will be a layer of iron-oxide passive film on the surface of steel bars in concrete members and it has a good protective effect. However, the protective passive film has been already neutralized when steel bars contract with carbonization deeply, and with the role of air and water, steel bars then begin to corrode and make the volume expand a little, which causes the protective film to crack and peel and reduces the strength of concrete. In addition, carbonization can also lead to contraction of concrete so that there will be tiny cracks on its surface. The advantage of carbonization is that the calcium carbonate generated by carbonization of concrete can fully fill the pores in cement paste to improve the density and prevent the intrusion of harmful substances.

The factors affecting carbonization are: the varieties of cement, the water-cement ratio, the external factors, mainly refer to the concentration of carbon dioxide and humidity in the air.

The measures to improve the anti-carbonization of concrete are to reduce water-cement ratio and mix water-reducing agent or air-entraining agent, both of which can enhance the density of concrete and its impermeability and slow down the carbonization rate.

5. Alkali-aggregate Reaction of Concrete

The alkali-aggregate reaction of concrete is that alkalis (Na_2O and K_2O) in cement react with active silicon dioxide in aggregate to generate alkali-silicic acid gel on the surface of aggregate. This gel has the property of wet swelling. When it swells, the cement paste surrounding aggregate expands and cracks. Such phenomenon that damages concrete is called alkali-aggregate reaction.

The reasons for the occurrence of alkali-aggregate reaction: first, the alkali content (the Na_2O content is over 0.6%) in cement is too high; second, there is active silica in aggregate; third, there is water in cement paste.

The main measures to avoid alkali-aggregate reaction: to adopt low-alkali cement, mix active blending materials, reduce swelling reaction, and mix with air-entraining agent and the aggregate without active Si_2O.

6. Measures to Improve Durability of Concrete

The major measures to improve the durability of concrete are listed in the following:

(1) Selecting cement appropriately.

(2) Controlling the maximum W/C and the minimum quantity of cement.

(3) Selecting appropriate aggregate, admixture, etc.

(4) Guaranteeing construction quality.

5.3.5 Creep Properties

Creep is defined as the gradual increase of strain over time under sustained load. Creep of concrete is a long-term process, and it takes place over many years. Although the amount of creep in concrete is relatively small, it could affect the performance of structures. The effect of creep varies with the type of structure. In simply supported reinforced concrete beams, creep increases the deflection, therefore, increases the stress in the steel. In reinforced concrete columns, creep results in a gradual transfer of load from the concrete to the steel. Creep also could result in losing some of the prestress in prestressed concrete structures, although the use of high-tensile stress steel reduces this effect.

5.4 The Quality Control of Concrete

The guarantee rates of quality of concrete directly impact the reliability and security of concrete structures and is the important aspects of modern scientific management.

The quality of concrete should be indicated by the result of testing its properties. In the actual projects, raw materials, construction and test conditions and other factors will affect the quality of concrete.

The influencing factors of raw materials and construction conditions are:

(1) The fluctuation of the quality and measuring of cement, aggregates, additives, and other raw materials.

(2) The fluctuation of water-cement ratio caused by the change of water consumption and the water content of aggregates.

(3) The fluctuation of stirring, transport, pouring, vibration, curing conditions and temperature changes.

The influencing factors of test conditions include: differences in sampling methods, specimen moulds, curing conditions, errors of testing machines, and the proficiency of laboratory personnel.

In the normal continuous production of concrete, mathematical statistical method can be adopted to inspect whether the strength of concrete or other technical indexes meet the quality requirements. The statistical methods include arithmetic mean, standard deviation, coefficient of variation, and the guarantee rate of strength which can comprehensively evaluate the quality of concrete. In the management of concrete production, the compressive strength of concrete is relevant to other properties well, so the compressive strength is usually employed to assess the quality of concrete in actual works.

5.5 The Design of the Mix Proportion of Concrete

The mix proportion of ordinary concrete refers to the proportion among the numbers of all the composed materials of concrete.

5.5.1 The Basic Points for the Mix Proportion Design

1. The Basic Requirements

(1) Strength: conformed with structure design.

(2) Feasibility: conformed with construction condition.

(3) Durability: conformed with engineering environment.

(4) Economical.

2. Principle of Mix Proportion Design of Concrete

The basic theory of mix proportion design is based on the change rule of concrete performance. Normal concrete mix proportion has four basic variables: C, W, S, G.

Three kinds of proportion:

(1) The ratio of water to cement;

(2) The ratio of sand to gravel;

(3) The quality of water in 1m^3 concrete ration of cement paste to aggregates.

3. Representation of the Mix Proportion Design

(1) It is represented by the masses of all the materials in each 1 m^3 concrete. For example, if there is 300 kg cement, 180 kg water, 720 kg sand, and 1200 kg stone, the total mass of each 1 m^3 concrete is 2400 kg.

(2) It is also represented by the mass proportion between all the materials. If the mass of cement counts 1, the above example can be converted into mass ratio: cement ∶ sand ∶ stone = 1 ∶ 2.4 ∶ 4, and water-cement ratio =0.60.

4. Tasks of Mix Proportion Design

(1) Picking out the appropriate raw material according to technical properties, structure and construction.

(2) Ascertaining the required technical economic index.

(3) Ascertaining the quantity of each material.

5.5.2 The Methods and Steps of the Mix Proportion Design

The design of mix proportion should be conducted in three steps: calculation of the preliminary mix proportion, the design of the proportion in labs, and the confirmation of the proportion in construction.

1. Calculation of the Preliminary Mix Proportion

(1) Determination of the Confected Strength ($f_{cu,0}$)

Based on the standard value of designed strength ($f_{cu,k}$) and the guarantee rate of 95%, the confected strength of concrete can be defined by the formula:

$$f_{cu,0} \geqslant f_{cu,k} + 1.645\sigma$$

In this formula: $f_{cu,0}$ is the produce strength of concrete (MPa);

$f_{cu,k}$ is the designed cubic standard compression strength of concrete;

σ is the standard deviation of concrete strength.

If statistical data is not available, σ can be calculated according to the following formula:

$$\sigma_0 = \sqrt{\frac{\sum_{i=1}^{n} f_{cu,i}^2 - n\overline{f_{cu}^2}}{n-1}}$$

If statistical data is not available, select σ according to the Table 5.10.

Table 5.10 Selection of σ

Concrete Strength Grade	Less Than C20	C20-C35	Higher Than C35
σ	4.0	5.0	6.0

(2) Selection of Water-cement Ratio(W/C)

① According to Strength Ratio

According to the actual strength of cement that has been tested f_{ce} (or the strength grade of cement f_{ce}^b), the aggregate types and the required confected strength of concrete $f_{cu,0}$, the water-cement ratio can be defined by the empirical formula of concrete strength:

$$f_{cu,0} = \alpha_a f_{ce}\left(\frac{C}{W} - \alpha_b\right)$$

Converted into:

$$\frac{W}{C} = \frac{\alpha_a f_{ce}}{f_{cu,0} + \alpha_a \alpha_b f_{ce}}$$

② According to Durability

To meet the demands of durability, the calculated W/C should not excess the value in Table 5.11.

Table 5.11 Maximum of W/C and minimum of cement

<table>
<tr><th rowspan="2" colspan="2">Environment Condition</th><th rowspan="2">Structure Type</th><th colspan="3">Maximum of W/C</th><th colspan="3">Minimum of Cement/kg</th></tr>
<tr><th>Concrete</th><th>Reinforced Concrete</th><th>Prestressed Concrete</th><th>Concrete</th><th>Reinforced Concrete</th><th>Prestressed Concrete</th></tr>
<tr><td colspan="2">Dry environment</td><td>Normal residences and offices</td><td>-</td><td>0.65</td><td>0.60</td><td>200</td><td>260</td><td>300</td></tr>
<tr><td rowspan="2">Damp environment</td><td>Without thawing</td><td>High humidity room; outside the house; in the earth and water (non-corrosive)</td><td>0.70</td><td>0.60</td><td>0.60</td><td>225</td><td>280</td><td>300</td></tr>
<tr><td>Thawing</td><td>Exterior members under thawing; member under thawing in the earth and water (non-corrosive); interior members under thawing in a high humidity</td><td>0.55</td><td>0.55</td><td>0.55</td><td>250</td><td>280</td><td>300</td></tr>
<tr><td colspan="2">Damp environment with thawing and deice</td><td>Exterior and interior members with thawing and deice</td><td>0.50</td><td>0.50</td><td>0.50</td><td>300</td><td>300</td><td>300</td></tr>
</table>

According to slump, coarse aggregate, maximum particle diameter and mixing water can be estimated from the Table 5.12.

Table 5.12 Mixed water in stiff concrete and plastic concrete

<table>
<tr><th colspan="2">Thickness of Concrete Composite</th><th colspan="3">Maximum Size of Gravel/mm</th><th colspan="3">Maximum Size of Crushed Stone/mm</th></tr>
<tr><td>Items</td><td>Indexes</td><td>10</td><td>20</td><td>40</td><td>16</td><td>20</td><td>40</td></tr>
<tr><td rowspan="3">Thickness</td><td>15-20</td><td>175</td><td>160</td><td>145</td><td>180</td><td>170</td><td>155</td></tr>
<tr><td>10-15</td><td>180</td><td>165</td><td>150</td><td>185</td><td>175</td><td>160</td></tr>
<tr><td>5-10</td><td>185</td><td>170</td><td>155</td><td>190</td><td>180</td><td>165</td></tr>
<tr><td rowspan="4">Slump /mm</td><td>10-30</td><td>190</td><td>170</td><td>150</td><td>200</td><td>185</td><td>165</td></tr>
<tr><td>30-50</td><td>200</td><td>180</td><td>160</td><td>210</td><td>195</td><td>175</td></tr>
<tr><td>50-70</td><td>210</td><td>190</td><td>170</td><td>220</td><td>205</td><td>185</td></tr>
<tr><td>70-90</td><td>215</td><td>195</td><td>175</td><td>230</td><td>215</td><td>195</td></tr>
</table>

(3) Calculation of Unit Cement Consumption of Concrete (m_{c0})

According to the cement quantity of $1m^3$ concrete and identified water-cement ratio (W/C), the cement consumption can be calculated by the following equation:

$$m_{c0} = m_{w0} / (W / C)$$

Based on the application conditions of structures and requirements of durability, the minimum cement consumption of 1 m^3 concrete can be found out in Table 5.11. Finally, the bigger value of the two results can be determined as the cement quantity of $1m^3$ concrete.

(4) Determination of Sand Percentage (β_s)

The reasonable sand percentage should be determined by slump, cohesion and water retention of concrete mixtures. Generally, the reasonable rate should be found out through experiments or based on the selection experiences of makers. In the absence of experience, the value can be selected from Table 5.13 on the basis of aggregate types and water-cement ratios.

Table 5.13 Selection of β_s

W/*C*	Maximum Size of Crushed Stone/mm			Maximum Size of Gravel/mm		
	16	20	40	10	20	40
0.40	30-35	29-34	27-32	26-32	25-31	24-30
0.50	33-38	32-37	30-35	30-35	29-34	28-33
0.60	36-41	35-40	33-38	33-38	32-37	31-36
0.70	39-44	38-43	36-41	36-41	35-40	34-39

(5) Calculation of Sand and Stone Consumption of 1 m^3 Concrete

The consumption of sand and stone can be obtained by mass method or volume method.

① Volume Method

Assumed that the volume of concrete mixtures equals to the sum of the absolute volume of all the components and the volume of the air contained in mixtures, the consumption of all the mixing materials can be identified as follows:

$$\beta_s = \frac{m_{s0}}{m_{s0} + m_{g0}} \times 100\%$$

$$\frac{m_{c0}}{\rho_c} + \frac{m_{w0}}{\rho_w} + \frac{m_{s0}}{\rho_s} + \frac{m_{g0}}{\rho_g} + 0.01\alpha = 1$$

In this formula: ρ_c and ρ_w are the densities of cement and water respectively (g/cm^3);

ρ_s and ρ_g are the apparent densities of sand and stone respectively (kg/m^3);

α is the percentage of air content in concrete; α=1 if air-entraining agent is not used.

Solve the above equation and can get m_{s0} and m_{g0}.

② Mass Method

Based on experience, if the state of raw materials is stable, the apparent density of prepared concrete mixtures should be approximate to a fixed value. Then an apparent density can be assumed at first, m_{s0} and m_{g0} can be calculated as follows:

$$m_{c0} + m_{g0} + m_{s0} + m_{w0} = m_{cp}$$

$$\frac{m_{s0}}{m_{s0} + m_{g0}} \times 100\% = \beta_s$$

In this formula: m_{cp} is the assumed apparent density of concrete mixtures (kg/m^3); m_{cp} value can be selected within 2350-2450 kg/m^3 in accordance with the apparent density and particle size of aggregates and the strength grades of concrete.

Through the above six-step calculation, the total consumptions of water, cement, sand and stone can be calculated to get the preliminary mix proportion. However, most of the above calculations are obtained by empirical formulas or resources. And the concrete prepared by those data may not meet the actual demands. Thus, the mix proportion should be tested, adjusted and identified.

2. Determination of the Lab Mix Proportion

(1) Adjustment of Workability

Based on the preliminary mix proportion, the amount required to prepare the concrete mixture is calculated as 15 L. At first, determine the slump through test and observe cohesion and water retention at the same time. It should be adjusted if inconsistent with the requirement. The principles of adjustment are as follows: if the mobility is too big, increase sand and stone appropriately, with the same sand ratio; if the mobility is too small, increase water and cement, water-cement ratio unchangeable; if cohesion and water retention are not good, sand ratio can be raised or reduced to adjust the workability in order to satisfy the required mix proportion, namely, the standard mix proportion for the test of concrete strength. When trial mix and adjusting work have been done, the actual apparent density of concrete mixtures should be tested ($\rho_{c,t}$).

(2) Rechecking of Strength

The water-cement ratio of the tested standard mix proportion may be not appropriate and the strength of concrete may not accord with requirements, so the concrete strength should be rechecked. Three different mixture ratios will be adopted in rechecking. One is the standard ratio, and the other two are the ratios whose water-cement ratios are increased and reduced by 0.05 respectively. The water consumptions of the other two ratios are the same with the one of the standard ratios, but their sand percentages are respectively increased and reduced by 1%. construct or calculate the strength and its corresponding water-cement ratio to obtain the confected strength ($f_{cu,0}$) and its water-cement ratio. Finally determine the consumptions of all the materials in 1 m^3 concrete by the following rules:

Water consumption (m_{w0}) should be determined based on the slump or Vebe consistency measured in the production of specimens and the water consumption of the standard mix proportion.

Cement consumption (m_{c0}) should be determined through multiplying water consumption by selected water-cement ratio.

Contents of fine and coarse aggregates (m_{g0} and m_{s0}) should be adjusted based on water-cement ratios and the content of fine and coarse aggregate of the standard mix proportion.

(3) Correction of the Apparent Density of Concrete

The mix proportion rechecked by strength should be corrected by the actually tested apparent density of concrete mixtures ($\rho_{c,t}$)to determine the consumptions of all the materials in 1 m^3 concrete. The steps are as follows:

First, calculate the apparent density of concrete mixtures ($\rho_{c,c}$):

$$m_c + m_g + m_s + m_w = \rho_{c,c}$$

Second, calculate the correction coefficient (δ):

$$\frac{\rho_{c,t}}{\rho_{c,c}} = \delta$$

Finally, calculate the lab mix proportion (the consumption of all the materials in 1 m^3 concrete) by the following formulas:

$$m_c = m_{cb}\delta$$
$$m_s = m_{sb}\delta$$
$$m_g = m_{gb}\delta$$
$$m_w = m_{wb}\delta$$

Specification for Mix Proportion Design of Ordinary Concrete (JGJ 55-2000) regulates: if the absolute value of the difference between apparent density's actual tested value and calculated value is no more than 2% of the calculated value, apparent density cannot be corrected.

Through the adjustment of every index, the corrected lab mix proportion is the designed proportion of concrete.

3. The Determination of Construction Mix Proportion of Concrete

In the lab mix proportion of concrete, sand and stone are calculated in their dry state (the water content of sand is less than 0.5%, and that of stone is less than 0.2%). In fact, sand and stone stored on the building site contain a certain amount of water, according to which, the weights of site materials are calculated and the water consumption should also be amended at the same time. The amended consumptions of all the materials in 1 m^3 concrete are known as the construction mix proportion.

Now, it is assumed that water ratio of sand on the building site is *a*% and that of stone is *b*%, and the consumptions of all the materials in 1 m^3 concrete are respectively m'_c, m'_g, m'_w, m'_s (kg). Then the weights of all the materials should be:

$$m'_{c} = m_{c}$$
$$m'_{s} = m_{s}(1+a\%)$$
$$m'_{g} = m_{g}(1+b\%)$$
$$m'_{w} = m_{w} - m_{s} \times a\% - m_{g} \times b\%$$

Questions

5.1 What are the advantages of the characteristics of concrete?

5.2 How to determine the type and grade of cement in concrete? What are the requirements for aggregate in concrete?

5.3 What is water-reducing agent? Discuss the effect of water-reducing agent on the quality of Portland cement.

5.4 Briefly describe the major influencing factors of workability and the impacts.

5.5 What is the basis for grading concrete? How to divide concrete grade?

5.6 How to calculate the preliminary mix proportion? How to calculate the lab mix proportion?

References

[1] HAIMEI ZHANG. Building Materials in Civil Engineering [M]. Beijing: Science Press, 2011.

[2] MICHAEL S MAMLOUK, JOHN P ZANIEWSKI. Materials for Civil and Construction Engineers[M]. Stockton: QWE Press, 2011.

Chapter 6
Masonry

6.1 Introduction

Masonry in the form of brickwork, blockwork and natural stone make it absolutely one of the most popular construction materials. Perhaps it is for this reason that little thought needs to be given to the design and building of masonry walls. This of course is far from the case as such walls have to perform simultaneously many functions including structure, thermal and sound insulation and weather protection as well as division of space. Each function has its own criteria to meet. Additionally, durability, cost and construction factors actually play important roles so that masonry wall design and construction is quite a complex project in fact.

The aim has been to treat each aspect in sufficient detail to enable the reader to appreciate the problems involved and their treatment in practice. The text, starting with a historical perspective, outlines current methods in design and construction whilst seeking to identify future trends.

6.2 Historical Perspective

Masonry is one of the oldest construction materials. Some ancient masonry structures enjoy world appreciate, such as the Pyramids of Egypt, the Great Wall of China, and Greek and Roman ruins. Masonry units are a popular construction material throughout the world and compete favorably with other materials, such as wood, steel, and concrete for certain applications.

In several millenniums, the archaeological and historical record show that the basic construction materials used to create human shelter were the same in the world and were derived from the earth or plant life. Figure 6.1 shows an example of a Stone Age house at

Figure 6.1 Sara Brae in Orkney

Skara Brae in the Orkney Islands.

The requirement of buildings and public facilities, as societies became more complex, progressed beyond the need of rudimentary shelter. This phenomenon gave a rise to the evolution of masonry construction from crude assemblages of small stones or stone slabs jointed with mud to large structures built with shaped blocks. Thus the Egyptian King Zoster had constructed the famous stepped pyramid at Sakkara by 2700 BC. Lots of other pyramids were built by the Pharaohs in the following centuries, without cranes, pulleys or lifting tackle but immense expenditure of labors and material. Numerous temples, with more complexity than pyramids, were also built by the Egyptians, which demonstrates a very high level of engineering skill in masonry construction at that time.

The Great Wall (Figure 6.2), one of the greatest vestiges of the world, was listed as a World Heritage by UNESCO in 1987. The Great Wall just like a gigantic dragon winds up and down across deserts, grasslands, mountains and plateaus, stretching approximately 8851.8 kilometers (5500 miles) from east to west of China. With a history of more than 2000 years from the Warring States Period (476 BC-221 BC) to Ming Dynasty (1368-1644), some of the sections are now in ruins or have disappeared. However, it is still one of the most appealing attractions all around the world owing to its architectural grandeur and historical significance. The mystery of the construction of the wall is amazing. The construction, which drew heavily on the local resources for construction materials, was carried out in-line with the local conditions under the management of contract and responsibility system. A great army of manpower, composed of soldiers, prisoners and local people, built the wall.

Figure 6.2 The Great Wall

The style of Roman building got continued in the Byzantine or Eastern Empire especially with the construction of churches, some on a very large scale such as Hagia Sophia in Constantinople (now Istanbul) in the 6th century AD. Through the most notable masonry buildings in Europe were churches, cathedrals and castles in the middle ages, the latter ranged from vast complexes with formidable defensive walls to relatively small towers in disputed areas. An example of a border tower is Borthwick Castle, near Edinburgh (Figure 6.3), built in 1430 with walls over 4 m thick.

In the 18th and 19th centuries the Industrial Revolution took place in Europe, the population expanded rapidly and cities grew proportionately. This resulted in building on a previously unprecedented scale. With the exception of some industrial building, this was almost entirely in masonry and timber until near the end of the 19th century. Some of the building in towns and cities was of a high standard but housing for the new working class was generally not and rapidly deteriorated into slums.

The earliest high rise buildings were built in masonry. One of the early skyscrapers in Chicago was the Monadnock Building (Figure 6.4), now preserved for its historic interest.

Extensive researches in many countries had led to the development of more sophisticated codes of practice, but masonry has been used almost entirely for low to medium rise buildings and for cladding of steel and concrete buildings in recent years, and the potential for high rise construction has not been fully explored. Figures 6.5 shows examples of late 20th century masonry buildings and the use of the materials cladding.

Figure 6.3 Borthwick Castle

Figure 6.4 The Monadnock Building

(a) St.Barnabas Church

(b) Offices in Sunderland

(c) Natural stone cladding

Figure 6.5 20th century masonry buildings

6.3 Masonry Units and Others

The classifications of the masonry materials, based on the appearances and properties, vary from different countries in the world. Here is one suggested by Professor Michael S.Mamlouk from America.

6.3.1 Masonry Units

Masonry units can be classified as:

(1) Concrete masonry units;

(2) Clay bricks;

(3) Structural clay tiles;

(4) Glass blocks;

(5) Stone.

It should also be noted that concrete masonry units can be either solid or hollow, but clay bricks, glass blocks, and stone are typically solid. Structural clay tiles are hollow units that are larger than clay bricks and are used for structural and non-structural masonry applications, such as partition walls and filler panels. They can be used with their webs in either a horizontal or a vertical direction. Figure 6.6 shows examples of concrete masonry units, clay bricks, and structural clay tiles.

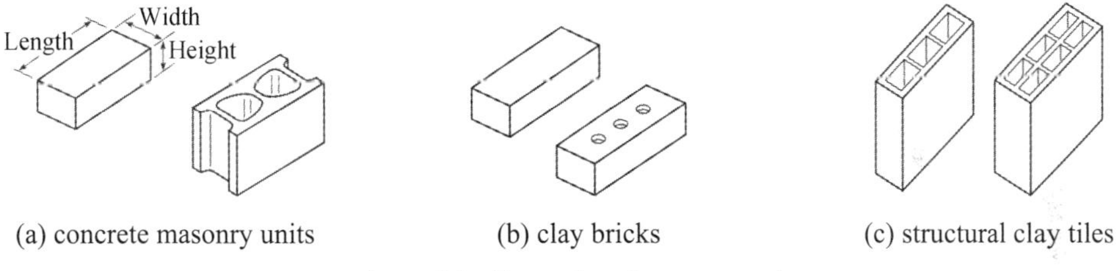

(a) concrete masonry units (b) clay bricks (c) structural clay tiles

Figure 6.6 Examples of masonry units

1. Concrete Masonry Units

Solid concrete units are usually named concrete bricks, but hollow units are known as concrete blocks, hollow blocks, or cinder blocks. Hollow units have a net cross-sectional area in every plane parallel to the bearing surface in Figure 6.7.

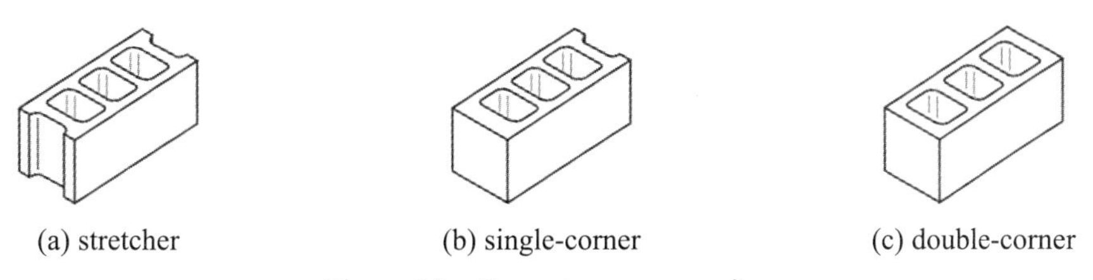

(a) stretcher (b) single-corner (c) double-corner

Figure 6.7 Concrete masonry units

There are three classes about concrete masonry units, based on their density: light-weight units, medium-weight units and normal-weight units. Well-graded sand, gravel, and crushed stone are used to manufacture normal-weight units. Light-weight aggregates such as pumice, scoria, cinders, and expanded clay. And on the other hand, expanded shale is used to manufacture light-weight units. The ability of light-weight units is that they have higher thermal and fire resistance properties and lower sound resistance than normal weight units.

Concrete masonry units are manufactured using a relatively dry concrete mixture consisting of Portland cement, aggregates, water, and admixtures. Air-entrained concrete is sometimes used to increase the resistance of the masonry structure to freeze and thaw effects and to improve workability, compaction, and molding characteristics of the units during manufacturing. The units are molded under pressure, then cured, usually using low-pressure steam curing. After manufacturing, the units are stored under controlled conditions so that the concrete continues curing.

Concrete masonry units are available in different sizes, colors, shapes, and textures. They are specified by their nominal dimensions. The nominal dimension is greater than its specified (or modular) dimension by the thickness of the mortar joint.

Solid concrete masonry units (concrete bricks) are manufactured in two types based on their exposure properties: concrete building bricks and concrete facing bricks. The concrete building bricks are manufactured for general use in non-facing, utilitarian applications, while the concrete facing bricks are typically used in applications where one or more faces of the unit is intended to be exposed. The concrete facing bricks have stricter requirements than the concrete building bricks. The maximum allowable water absorption of the concrete facing bricks is less than that of the concrete building bricks. Also, the minimum net area compressive strength of the concrete facing bricks is higher than that of the concrete building bricks.

2. Clay Bricks

As we known, clay bricks are used for different applications, including building, facing and aesthetics, floor making, and paving. Building bricks (common bricks) are used as a structural material, and are typically strong and durable. Facing bricks are used for facing and aesthetic purposes, and are available in different sizes, colors, and textures. Floor bricks are used on finished floor surfaces, and are generally smooth and dense, with high resistance to abrasion. Finally, paving bricks are used as a paving material for roads, sidewalks, patios, driveways, and interior floors. Paving bricks are available in different colors, such as red, gray, or brown.They are typically abrasion resistant, and could usually be vitrified (glazed to render it impervious to water and highly resistant to corrosion).

How to define clay bricks and produce them? Clay bricks are small, rectangular blocks made of fired clay. And the clays used for brick making vary widely in composition from one place to another. On one hand, clays are composed mainly of silica (grains of sand), alumina, lime, iron, manganese, sulfur, and phosphates, with different proportions. On the other hand, bricks are manufactured by grinding or crushing the clay in mills and mixing it with water to make it plastic. The plastic clay is then molded, textured, dried, and finally fired. Bricks are manufactured in different colors, such as dark red, purple, brown, gray, pink, or dull brown, depending on the firing temperature of the clay during manufacturing.

Clay bricks have some satisfied performances, what we should know is that absorption is one of the important properties that determine the durability of bricks. Highly absorptive bricks can cause efflorescence and other problems in the masonry. And clay bricks have strength enough for low building, they are very durable and fire resistant, and require very little maintenance. They have moderate insulating properties, which make brick houses cooler in summer and warmer in winter, compared with houses built with other construction materials. Clay bricks are also noncombustible and poor conductors.

6.3.2 Other Materials

1. Mortar

Mortar classified as cement-lime mortar, cement mortar, or masonry cement mortar is an important material of masonry, it consists in cementitious material, aggregate, and water. Mortar is used for the following functions:

(1) Bonding masonry units together, either non-reinforced or reinforced;

(2) Serving as a seating material for the units;

(3) Leveling and seating the units;

(4) Providing aesthetic quality of the structure.

Mortar needs to satisfy either proportion specifications or property specifications. The proportion specifications specify the ingredient quantities, while the property specifications specify the compressive strength, water retention, air content, and the aggregate ratio.

Mortar starts to bind masonry units when it sets. During construction, bricks and blocks should be rubbed and pressed down in order to force the mortar into the pores of the masonry units to produce maximum adhesion. It should be noted, however, that mortar is the weakest part of the masonry wall. Therefore, thin mortar layers generally produce stronger walls than do thick layers.

Unlike concrete, the compressive strength is not the most important property of mortar. Since mortar is used as an adhesive and sealant, it is very important that it forms a complete, strong, and durable bond with the masonry units and with the bars that might be used to reinforce masonry walls. The ability to bond individual units is measured by the tensile bond strength of mortar, which is related to the force required to separate the units. Other properties that affect the performance of mortar are workability, tensile strength, compressive strength, resistance to freeze and thaw, and water retention.

2. Grout

Grout, which is a high-slump concrete consisting of Portland cement, sand, fine gravel, water, and sometimes lime, is used to fill the cores or voids in hollow masonry units for the purpose of bonding the masonry units, bonding there in forcing steel to the masonry, increasing the bearing area and fire resistance, and improving the overturning resistance by increasing the weight.

3. Plaster

Plaster, which is a fluid mixture of Portland cement, lime, sand, and water, is used for finishing either masonry walls or framed (wood) walls, and for exterior or interior walls. Stucco is plaster used to cover exterior walls.

6.4 Advantages and Disadvantages

6.4.1 Advantages

Masonry materials are available with properties capable of meeting all varied requirements, requiring only to be supplemented by the use of other materials for thermal insulation, damp-proof courses, wall ties and the like.

The first advantage of masonry lies in the fact that a single element can fulfill several functions thus leading to simplified and economical construction. These functions can include structure, fire protection, sound insulation, thermal insulation, weather exclusion and sub-division of space.

The second major advantage relates to the durability of masonry materials. As long as the selection of masonry materials is appropriat, it can be expected to remain serviceable for many decades, if not centuries, with relatively little maintenance.

Besides, from the architectural point of view, masonry offers great flexibility in terms of plan form, spatial composition and facing for which materials are available in an almost infinite variety of colors and textures. These features offer great flexibility in design, as has been demonstrated in buildings of many different categories.

6.4.2 Disadvantages

As we all known, every coin has two sides, the masonry also has some bothering breaches as below:

(1) Extreme weather, under certain circumstances, can cause degradation of masonry

wall surfaces due to frost damage.

(2) Masonry tends to be heavy and must be built upon a strong foundation, such as reinforced concrete, to avoid settling and cracking.

(3) Other than concrete, masonry construction does not lend itself well to mechanization, and requires more skilled labor than stick-framing.

6.5 Future Trends

There are still some researches needed in masonry structural materials: the study of high-strength material (high-strength blocks and blocks type), the basic material of high performance mortar and grouted concrete.

Besides, for masonry to retain its place as a primary construction material, particularly as a cladding, there will be a need to improve building techniques with a view to reducing construction time on site. The means of achieving this may be expected to include the use of larger units requiring the use of mechanical handling in placing them on the wall. Such units will be dimensionally accurate leading to thinner jointing and the use of smaller quantities of mortar, possibly of new types. Future use of off-site construction is likely to be limited to smaller elements, which would not call for highly expensive factory plant, transportation and lifting equipment.

In addition to these well-known trends there has recently entered into prominence the need to take account of "sustainability" and possible climate change in decisions relating to building design, particularly in response to the now generally accepted phenomenon of global warming.

Questions

6.1 What are the advantages of masonry?

6.2 How to classify concrete masonry units based on their density and based on their exposure properties?

6.3 How to define clay bricks and produce them?

6.4 What is Mortar? Introduce its functions and application.

6.5 According to the advantages and disadvantages of masonry, offer some suggestions for improvement.

References

[1] KHALAF F M. Masonry Wall Construction [M]. Memphis: Henry Dream Press, 2001.

[2] MICHAEL S MAMLOUK, JOHN P ZANIEWSKI. Materials for Civil and Construction Engineers[M]. Stockton: QWE Press, 2011.

[3] Travel China Guide. Great Wall of China [OL]. [2020-11-06] http://www.travelchinaguide.com/china_great_wall/.

[4] 蓝宗建. 砌体结构[M]. 北京：中国建筑工业出版社，2013.

Chapter 7

Asphalt and Asphalt Concrete

7.1 Introduction

Asphalt is one of the oldest materials used in construction. The first recorded use of asphalt dates back to 3800 BC in Mesopotamia where the material was used as an adhesive mortar for building stones and paving blocks. Asphalt was first used in paving in Babylon around 625 BC, in the reign of King Naboppolassar.

Asphalt, also known as bitumen is a sticky, black and highly viscous liquid or semi-solid form of petroleum. It may be found in natural deposits or may be a refined product. The materials used by these engineers of ancient times were naturally occurring asphalts. Such materials are still used today, often for similar purposes. However, the vast bulk of today's bituminous materials are the products of industrial refining processes.

7.2 Classification

Bituminous materials are classified as asphalts and tars, as shown in Figure 7.1. Asphalt is used mostly in pavement construction, but is also used as sealing and waterproofing agents.

Tar is produced by the destructive distillation of bituminous coal or by cracking petroleum vapors. Tar is used primarily for waterproofing membranes, such as roofs. Tar may also be used for pavement treatments, particularly where fuel spills may dissolve asphalt cement, such as on parking lots and airport aprons.

7.3 Asphalt Cement

7.3.1 Composition

Asphalt cement is a mixture of a wide variety of hydrocarbons primarily consisting of hydrogen and carbon atoms, with minor components such as sulfur, nitrogen, oxygen (heteroatoms), and trace metals. The percentages of the chemical components, as well as the molecular structure of asphalt, vary depending on the crude oil source.

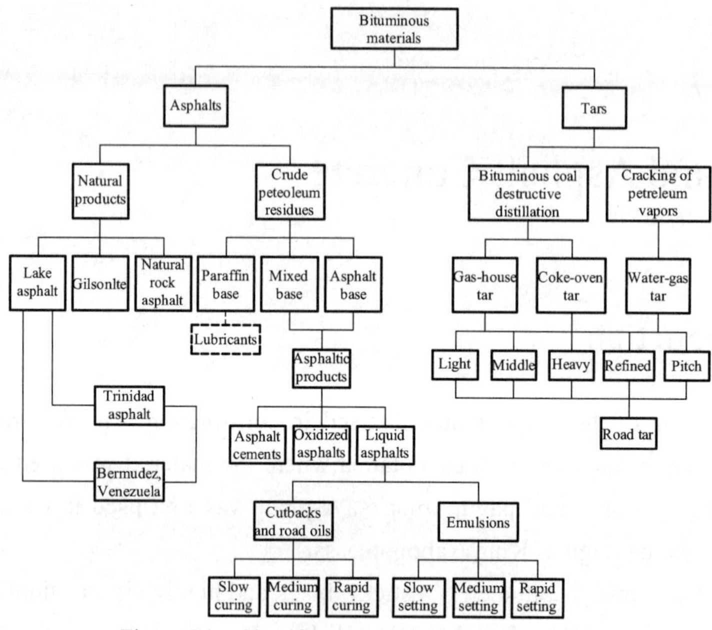

Figure 7.1 Classification of bituminous materials

Because asphalt contains such a large number of molecules with differing chemical composition and structure, complete chemical analysis is generally impractical. In general, asphalt can be divided into two broad chemical groups: asphaltenes and maltenes. The maltenes can be further subdivided into saturates, aromatics and resins. Chromatographic techniques are the most common methods to separate asphalt into four fractions on the basis of the size, chemical reactivity and polarity. These four fractions are often called SARA fractions (Saturates, Aromatics. Resins and Asphaltenes). The characteristics of four fractions of asphalt are shown in Table 7.1.

Table 7.1 Characteristics of four fractions of asphalt

Component	Color	State	Molecular weight	Contribution
Saturates	Transparent	Liquid	-	-
Aromatics	Pale yellow	Liquid	300-500	Viscosity and fluidity
Resins	Yellow	Semisolid	600-1000	Adhesion and ductility
Asphaltenes	Dark brown	Solid	-	Strength and stiffness

7.3.2 Colloidal Structure of Asphalt

The asphaltenes with high molecular weight are wrapped by resins, which form micelles and are dispersed in oil media. This kind of colloidal system has a unique liquidity and typical rheological properties. The colloidal structure varies due to different content of each composition.

1. Sol Structure

The quantities of resin/aromatic fraction are adequate to make the asphaltenes peptized, and the attraction among the micelles is low. This kind of asphalt is characterized by the great fluidity but inferior viscosity. The straight-run asphalt is of the sol structure.

2. Gel Structure

If the quantity of the aromatic/resin fraction is insufficient to peptize the asphaltenes, the asphaltenes can associate to form large micelles or even a continuous network throughout the asphalt. This kind of asphalt is characterized by the inferior fluidity, plasticity and temperature sensitivity, but of great viscosity. The oxidized asphalt has the gel structure.

3. Sol-gel Structure

The asphalt contains appropriate ground asphaltenes and micelles, and the distances among micelles are relatively small, and there is certain attraction among them, and the structure between sol structure and gel structure is formed, called sol-gel structure. In practice, most asphalt is of intermediate characteristics.

The structure of petroleum asphalt depends not only on the content of the composition but also on the temperature.

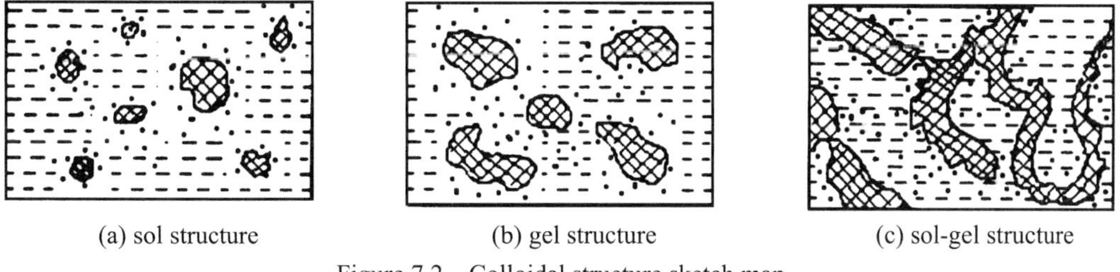

(a) sol structure (b) gel structure (c) sol-gel structure

Figure 7.2 Colloidal structure sketch map

7.3.3 Physical Properties

1. Density

The density of asphalt is the mass per unit volume at the given temperature. Generally, the value is between 0.97-1.01g/cm^3. The density is related to the chemical composition of asphalt.

2. Stiffness

At low temperatures, and in particular for short duration of loading, asphalt binder shows elastic properties, deforming under load and recovering on the removal of load. If loads are applied for an appreciable time, however, viscous flow develops. It has been demonstrated for asphalt binder that the rate of viscous flow decreases with time. This would not be the case for true liquids, and it therefore follows that this apparently viscous flow is made up of two components, a true viscous flow as in liquids, together with a delayed elastic deformation.

3. Viscosity

Viscosity is the fundamental material property relating the rate of shear stain in a fluid to the applied shear stress; in practical terms, it is the resistance to flow. The rheological behavior of asphalt cements is very temperature dependent, and therefore the temperature of a material must be stated when reporting its rheological properties.

4. Thixotropy

The property of asphalt binder whereby it "sets" when not stirred. Thixotropy is thought to result from hydrophilic suspended particles that form a lattice structure throughout the asphalt binder. This causes the asphalt to increase in viscosity and hardening. Thixotropic effects can be somewhat reversed by heat and agitation. The property is most marked in blown or oxidized asphalts, although it is exhibited to some extent by many binders.①

5. Softening Point

Softening point of asphalt is taken as the temperature at which the sample soft enough to allow the ball, enveloped in the sample material, to fall a distance of 25 mm. Asphalt binder does not change from solid to liquid at any definite temperature. It gradually become softer and less viscous as the temperature rises. Softening point is useful in the classification of certain asphalt binder. It is indicative of the tendency of the material to flow at elevated temperatures encountered in service.

6. Penetration

Asphalt binder is classified by the depth to which a standard needle will penetrate under specified test conditions. This "pen" test classification is used to indicate the hardness of asphalt, lower penetration indicating a harder asphalt. Specifications for penetration graded asphalts normally state the penetration range for a grade, e.g. 50/70.

7. Temperature Susceptibility

The consistency of asphalt is greatly affected by temperature. Asphalt gets hard and brittle at low temperatures and soft at high temperatures. The viscosity of the asphalt decreases when the temperature increases. Asphalt's temperature susceptibility can be represented by the slope of the line shown in Figure 7.3; the steeper the slope, the higher the temperature susceptibility of the asphalt. However, additives can be used to reduce this susceptibility.

---

① Pavement interactive. Asphalt Durability [OL]. [2020-01-01] http://www.pavementinteractive.org/article/durability/.

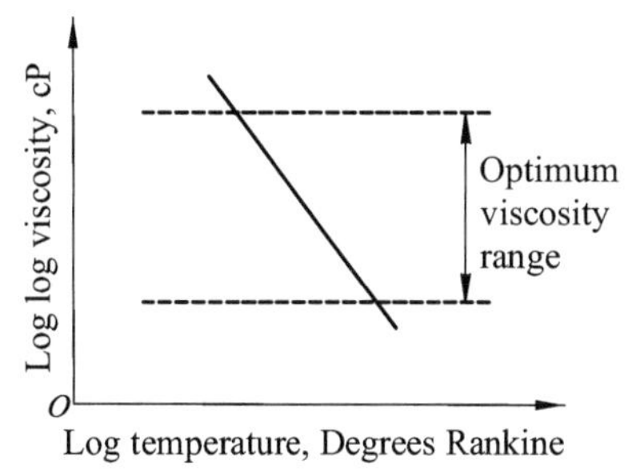

Figure 7.3 Typical relations between asphalt viscosity and temperature

Due to temperature susceptibility, the grade of the asphalt cement should be selected according to the climate of the area. The viscosity of the asphalt should be mostly within the optimum range for the area's annual temperature range; soft-grade asphalts are used for cold climates and hard-grade asphalts for hot climates.

8. Penetration Index

The penetration index represents a quantitative measure of the response of bitumen to variation in temperature. Knowing the penetration index of particular bitumen, it is possible to predict its behavior in an application. Therefore, asphalt binders with high penetration numbers (called "soft") are used for cold climates while asphalt binders with low penetration numbers (called "hard") are used for warm climates. All bitumen displays thermoplastic properties i.e. they become softer when heated and harden when cooled. Several equations exist that define the way that the viscosity (or consistency) changes with temperature. One of the best known is that developed by Pfeiffer and Van Doormaal which states that: if the logarithm of penetration, P, is plotted against temperature, T, a straight line is obtained such that:

$$\lg P = AT + K$$

In this formula: P is the penetration at temperature T;

A is temperature susceptibility (or temperature sensitivity);

K is the constant.

A penetration index (PI) has been defined for which the temperature susceptibility would assume a value of zero for road asphalt, as given by:

$$PI = 20(1 - 25A)/(1+50A)$$

The value of A (and PI) can be derived from penetration measurements at two temperatures, T_1 and T_2, using the equation:

$$A = [\lg P_{T_1,m,t} - \lg P_{T_2,m,t}]/(T_1 - T_2)$$

In this formula: m is the mass of the test sample;

t is the set time of the test.

7.3.4 Asphalt Grading Methods

1. Penetration Grading

Penetration grading's basic assumption is that the less viscous the asphalt, the deeper the needle will penetrate. This penetration depth is empirically (albeit only roughly) correlated with asphalt binder performance. Therefore, asphalt binders with high penetration numbers (called "soft") are used for cold climates while asphalt binders with low penetration numbers (called "hard") are used for warm climates. Table 7.2 shows the penetration grading system of asphalt cement.

Table 7.2 Penetration grading system of asphalt cement

Grade	Penetration min	Penetration max	Comments
40-50	40	50	Hard grade
60-70	60	70	Typical grades used on pavement
85-100	85	100	
120-150	120	150	Soft grade. Used for cold climates
200-300	200	300	

2. Viscosity Grading

Viscosity grading can be done on original (as-supplied) asphalt binder samples (called AC grading) or aged residue samples (called AR grading). The AR viscosity test is based on the viscosity of aged residue from the rolling thin film oven test. With AC grading, the asphalt binder is characterized by the properties it possesses before it undergoes the HMA manufacturing process. The AR grading system is an attempt to simulate asphalt binder properties after it undergoes a typical HMA manufacturing process and thus, it should be more representative of how asphalt binder behaves in HMA pavements. Table 7.3 and Table 7.4 show standard viscosity grades for the AC and AR grading systems (ASTM D3381).

Table 7.3 Standard viscosity grades based on original asphalt (AC)

Grading	AC-2.5	AC-5	AC-10	AC-20	AC-30	AC-40
Absolute Viscosity(Poises)	250±50	500±100	1000±200	2000±400	3000±600	4000±800

Table 7.4 Standard viscosity grades based on aged-residue asphalt(AR)

Grading	AR-1000	AR-2000	AR-4000	AR-8000	AR-16000
Absolute Viscosity(Poises)	1000±250	2000±500	4000±1 000	8000±2000	16000±4000

3. Superpave Performance Grading

Penetration grading and viscosity grading are somewhat limited in their ability to fully characterize asphalt binder for use in HMA pavement. Therefore, as part of the Superpave research, new binder tests and specifications were developed to more accurately and fully characterize asphalt binders for use in HMA pavements. These tests and specifications are specifically designed to address HMA pavement performance parameters such as rutting, fatigue cracking and thermal cracking.

Superpave performance grading (PG) is based on the idea that an HMA asphalt binder's properties should be related to the conditions under which it is used. For asphalt binders, this involves expected climatic conditions as well as aging considerations. Therefore, the PG system uses a common battery of tests (as the older penetration and viscosity grading systems do) but specifies that a particular asphalt binder must pass these tests at specific temperatures that are dependent upon the specific climatic conditions in the area of use. Table 7.5 shows how the Superpave PG system addresses specific penetration, AC and AR grading system general limitations.

Table 7.5 Prior limitations vs Superpave testing and specification features

Limitations of Penetration, AC and AR Grading Systems	Superpave Binder Testing and Specification Features that Address Prior Limitations
Superpave performance grading (PG) is based on the idea that an HMA asphalt binder's properties should be related to the conditions under which it is used. For asphalt binders, this involves expected climatic conditions as well as aging considerations. Therefore, the PG system uses a limitations of penetration, AC and AR grading systems	Superpave binder testing and Specification features that address prior limitations
Penetration and ductility tests are empirical and not directly related to HMA pavement performance	The physical properties measured are directly related to field performance by engineering principles
Tests are conducted at one standard temperature without regard to the climate in which the asphalt binder will be used	Test criteria remain constant, however, the temperature at which the criteria must be met changes in consideration of the binder grade selected for the prevalent climatic conditions
The range of pavement temperatures at any one site is not adequately covered. For example, there is no test method for asphalt binder stiffness at low temperatures to control thermal cracking	The entire range of pavement temperatures experienced at a particular site is covered

Continued

Limitations of penetration, AC and AR grading systems	Superpave binder testing and Specification features that address prior limitations
Test methods only consider short-term asphalt binder aging (thin film oven test) although long-term aging is a significant factor in fatigue cracking and low temperature cracking	Three critical binder ages are simulated and tested: (1) Original asphalt binder prior to mixing with aggregate. (2) Aged asphalt binder after HMA production and construction. (3) Long-term aged binder
Asphalt binders can have significantly different characteristics within the same grading category	Grading is more precise and there is less overlap between grades
Modified asphalt binders are not suited for these grading systems	Tests and specifications are intended for asphalt "binders" to include both modified and unmodified asphalt cements

Superpave performance grading uses the following asphalt binder tests: rolling thin film oven (RTFO), pressure aging vessel (PAV), rotational viscometer (RV), dynamic shear rheometer (DSR), bending beam rheometer (BBR), direct tension tester (DTT).

(1) Rolling Thin Film Oven (RTFO)

The rolling thin-film oven (RTFO) procedure is used to simulate the short-term aging that occurs in the asphalt during production of asphalt concrete. In the RTFO method (ASTM D2872), the asphalt binder is poured into special bottles. The bottles are placed in a rack in a forced-draft oven, at a temperature of 163°C (325°F) for 75 min. The rack rotates vertically, continuously exposing fresh asphalt. The binder in the rotating bottles is also subjected to an air jet to speed up the aging process. The performance grade specifications limit the amount of mass loss during RTFO conditioning. Rolling thin-film oven conditioning is used to prepare samples for evaluation for rutting potential with the dynamic shear rheometer and prior to conditioning with the pressure aging vessel. Under the penetration and viscosity grading methods, the aged binder is usually tested for penetration or viscosity and the results are compared with those of new asphalt.

(2) Pressure Aging Vessel (PAV)

The pressure aging vessel (PAV) provides simulated long term aged asphalt binder for physical property testing. Asphalt binder is exposed to heat and pressure to simulate in-service aging over a 7-10 years' period. According to ASTM D6521, the basic PAV procedure takes RTFO aged asphalt binder samples, places them in stainless steel pans and then ages them for 20 hours in a heated vessel pressurized to 305 psi (2.10 MPa or 20.7 atmospheres). Samples are then stored for use in physical property tests.

(3) Rotational Viscometer (RV)

The rotational (Brookfield) viscometer test (ASTM D4402) consists of a rotational coaxial cylinder viscometer and a temperature control unit. The test is used to determine the viscosity of unaged asphalt binders. The RV test can be conducted at various temperatures, but for Superpave PG asphalt binder, specification is always conducted at 275°F (135°C).

(4) Dynamic Shear Rheometer Test (DSR)

The dynamic shear rheometer is used to measure three specification requirements in the performance grading system. For testing the neat binder and for the rutting potential test, the test temperature is equal to the upper temperature for the grade of the asphalt binder (e.g., a PG 64-PG22 is tested at 64 °C). For these tests, the sample size is 25 mm in diameter and 1 mm thick. Prior to testing for rutting potential, the sample is conditioned in the rolling thin-film oven. For evaluating fatigue potential, the intermediate temperature is used; 25 °C for PG64-PG22. The sample size is 8 mm in diameter by 2 mm thick. Prior to testing, the sample is conditioned in the rolling thin-film oven, followed by the pressure-aging vessel.

The DSR measures a specimen's complex shear modulus (G*) and phase angle (δ). The complex shear modulus (G*) can be considered the sample's total resistance to deformation when repeatedly sheared, while the phase angle (δ), is the lag between the applied shear stress and the resulting shear strain. The larger the phase angle (δ) is, the more viscous the material is.

(5) Bending Beam Rheometer Test (BBR)

The bending beam rheometer (BBR) test provides a measure of low temperature stiffness and relaxation properties of asphalt binders. These parameters give an indication of an asphalt binder's ability to resist low temperature cracking. The BBR is used in combination with the direct tension tester (DTT) to determine an asphalt binder's low temperature PG grade. As with other superpave binder tests, the actual temperatures anticipated in the area where the asphalt binder will be placed determine the test temperatures used. The basic BBR test uses a small asphalt beam that is simply supported and emersed in a cold liquid bath. A load is applied to the center of the beam and its deflection is measured against time. Stiffness is calculated based on measured deflection and standard beam properties and a measure of how the asphalt binder relaxes the load induced stresses is also measured. BBR tests are conducted on PAV aged asphalt binder samples.

(6) Direct Tension Tester (DTT)

The direct tension tester (DTT) test provides a measure of low temperature stiffness and relaxation properties of asphalt binders. These parameters give an indication of an asphalt binder's ability to resist low temperature cracking. The DTT is used in combination with the BBR to determine an asphalt binder's low temperature PG grade determination.

The basic DTT test measures the stress and strain at failure of a specimen of asphalt binder pulled apart at a constant rate of elongation. Test temperatures are such that the failure will be from brittle or brittle-ductile fracture. The test is of little use at temperatures where the specimen fails by ductile failure (stretches without breaking). DTT tests are conducted on PAV aged asphalt binder samples.

7.4 Cutback Asphalt

Cutback asphalts is produced by mixing asphalt cement with a petroleum diluent. Cutback asphalt is classified into three groups depending on the relative speed of evaporation of the petroleum diluent: rapid curing (RC), medium curing (MC), slow curing (SC). Cutbacks have viscosities that are typically lower than that of neat asphalt. After a cutback is applied, a portion of the solvent will eventually evaporate leaving only the asphalt cement after it has cured. Their uses include dust control, pavement maintenance, surface treatment (chip seal), cold mix patching and so on.

7.5 Emulsified Asphalt

Emulsified asphalt is produced by dispersing tiny globules of asphalt cement into water treated with a small quantity of emulsifying agent. The water comprises the continuous phase and the globules of asphalt cement make up the discontinuous phase. The emulsion setting rate will determine the different grades for emulsified asphalt: rapid setting (RS), medium setting (MS), slow setting (SS), quick setting (QS).

The main applications of asphalt products include: cold mix patching, cold in place recycling, surface treatment (fog, sand, chip, sandwich, slurry seals, micro surfacing and cape seal), pavement maintenance and so on.

Asphalt emulsions are replacing cutbacks due to environmental concerns. The following four reasons that asphalt emulsions should be used in lieu of cutbacks.

(1) Environmental regulations. Emulsions are relatively pollution free. Unlike cutback asphalts, there are relatively small amounts of volatiles to evaporate into the atmosphere other than water.

(2) Loss of high energy products. When cutback asphalts cure, the diluents which are high energy, high price products are wasted into the atmosphere.

(3) Safety. Emulsions are safe to use. There is little danger of fire as compared to cutback asphalts, some of which have very low flash points.

(4) Lower application temperature. Emulsions can be applied at relatively low temperatures compared to cutback asphalt, thus saving fuel costs. Emulsions can also be applied effectively to a

damp pavement, whereas dry conditions are required for cutback asphalts.

7.6 Other Asphalt

1. Regenerating Asphalt

Regenerating asphalt comprises reclaimed asphalt and a rejuvenating agent. The reclaimed asphalt comprises aggregate and an aged binder. The rejuvenating agent, which has a cyclic content of at least 5 wt. %, comprises an ester or ester blend derived from an acid selected from aromatic acids, fatty acids, fatty acid monomers, fatty acid dimers, fatty acid trimers, rosin acids, rosin acid dimers, and mixtures thereof. The rejuvenating agent revitalizes the aged bitumen binder of reclaimed asphalt and restores its physical properties to those of the original performance-grade bitumen. Improvements include desirable softening, low-temperature cracking resistance, better fatigue cracking resistance, good elevated temperature performance, improved miscibility, and restored temperature sensitivity. The rejuvenating agents enable the use of higher levels of recovered asphalt, particularly RAP, in asphalt mixtures, reduce binder and aggregate costs, and help the construction industry reduce its reliance on virgin, non-renewable materials.

2. Biological Asphalt

When adding bio-oil to the conventional asphalt, this asphalt can be called bio-asphalt. Bio-oil is created by a thermochemical process called fast pyrolysis. Corn stalks, wood wastes or other types of biomass are quickly heated without oxygen. The process produces a liquid bio-oil that can be used to manufacture fuels, chemicals and asphalt plus a solid product called biochar that can be used to enrich soils and remove greenhouses gases from the atmosphere.

7.7 Asphalt Concrete

Asphalt concrete is a composite material commonly used in construction of roads, highways, airports, parking lots, and many other types of pavement. It is commonly called simply asphalt or blacktop. The terms "asphalt concrete", "bituminous asphalt concrete" and the abbreviation "AC" are typically used only in engineering and construction documents and technical literature where the definition of "concrete" is any composite material composed of mineral aggregate glued together with a binder, whether that binder is Portland cement, asphalt or even epoxy.①

---

① Virginia asphalt association. Asphalt Concrete [OL]. [2020-01-01] http://www.vaasphalt.org/asphalt-concrete/.

7.7.1 Classification

Asphalt concrete can be divided into three primary disciplines in accordance with the composition of the aggregate gradation.

1. Dense Graded Asphalt Mixtures

Dense-graded mixes are produced with well or continuously graded aggregate (gradation curve does not have any abrupt slope change) and intended for general use. Typically, larger aggregates "float" in a matrix of mastic composed of asphalt cement and screenings/fines. When properly designed and constructed, a dense-graded mix is relatively impermeable. Dense-graded mixes are generally referred to by their nominal maximum aggregate size. They can further be classified as either fine-graded or coarse-graded. Fine-graded mixes have more fine and sand sized particles than coarse-graded mixes.

2. Open Graded Asphalt Mixtures

Open graded asphalt mixes with relatively uniform-sized aggregate typified by an absence of intermediate-sized particles (gradation curve has a nearly vertical drop in intermediate size range). Mixes typical of this structure are the permeable friction course, generally called "Open Graded Friction Course" (OGFC) and asphalt-treated permeable bases. Because of their open structure, precautions are taken to minimize asphalt drain-down by the use of fibers and/or modified binders. Stone-on-stone contact with a heavy asphalt cement particle coating typifies these mixes.

3. Gap Graded Asphalt Mixtures

Gap-graded mixes use an aggregate gradation with particles ranging from coarse to fine with some intermediate sizes missing or present in small amounts. The gradation curve may have a "flat" region denoting the absence of a particle size or a steep slope denoting small quantities of these intermediate aggregate sizes. These mixes are also typified by stone-on-stone contact and can be more permeable than dense-graded mixes or highly impermeable, as in the case of stone matrix asphalt (SMA).

Asphalt concrete can be also divided into three categories according to the mixing and construction temperature.

4. Hot Mix Asphalt (HMA)

Hot mixes are produced at a temperature between 150 °C and 190 °C.

5. Warm Mix Asphalt (WMA)

A typical WMA is produced at a temperature around 20-40 °C lower than an equivalent hot mix asphalt. Less energy is involved and during the paving operations, the temperature in the mix is lower, resulting in improved working conditions for the crew and an earlier opening of the road.

6. Cold Mix

Cold mixes are produced without heating the aggregate. This is only possible, due to the use of a specific bitumen emulsion which breaks either during compaction or during mixing. After breaking, the emulsion coats the aggregate and over time, increases its strengths. Cold mixes are particularly recommendable for lightly trafficked roads.

7.7.2 Composite and Structure

Asphalt concrete is a composite material, consisting of two components or phases, an asphalt binder and an aggregate, which are physically combined. It can be divided into three types in accordance with the aggregate skeletal structure. There are more fine aggregates and fewer course aggregates.

1. Suspension Dense Structure

The aggregate size of this type is continuous. The coarse aggregates are separated by the fine aggregates and suspended between the fine aggregates and asphalt mortar. This structure has high density, good water stability and durability, but low high-temperature stability. Dense-graded asphalt concrete has this structure.

2. Skeleton Gap Structure

This type has more coarse-aggregates and less fine aggregates than the suspension dense structure type. The coarse aggregates connect with each other forming skeleton gap structure, thus more voids exist in this structure. This type of asphalt concrete has good high-temperature stability, poor water stability and durability. Representative of this type is open-graded friction course (OGFC).

3. Dense Skeleton Structure

The coarse aggregates can form skeleton structure and the voids among the skeleton are filled with fine aggregates. The high-temperature stability, water stability and durability of this type of asphalt concrete are all good, but it is easy to suffer from segregation. Stone mastic asphalt (SMA) is the typical representative of this structure.

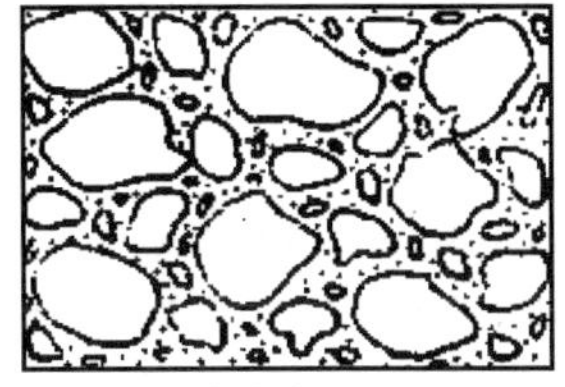

(a) suspended dense structure

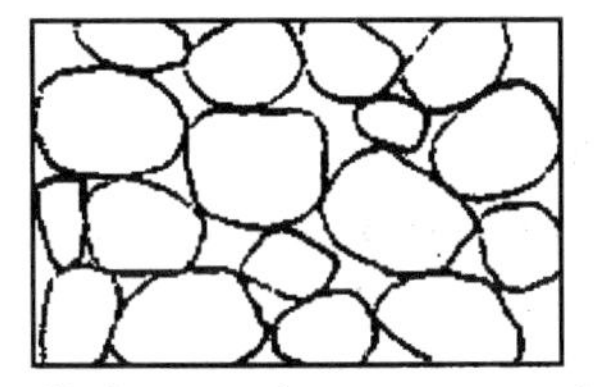

(b) framework-pore structural

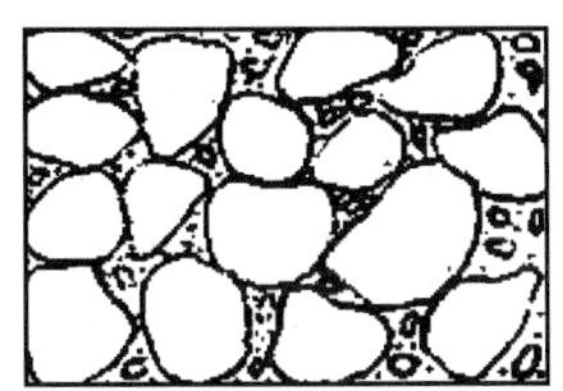

(c) dense skeleton structure

Figure 7.4 Aggregate skeletal structure

7.7.3 Mix Design

1. Marshall Mix Design

The basic concepts of the Marshall mix design method were originally developed by Bruce Marshall of the Mississippi Highway Department around 1939 and then refined by the U.S. Army. Currently, the Marshall method is widely used in the world. The Marshall method seeks to select the asphalt binder content at a desired density that satisfies minimum stability and range of flow values.

The Marshall mix design method consists of 6 basic steps: (1) aggregate selection, (2) asphalt binder evaluation, (3) sample preparation (including compaction), (4) stability determination, (5) density and voids analysis, (6) optimum asphalt binder content selection.

(1) Aggregate Selection

The aggregate physical properties which consist of toughness, abrasion, durability, soundness, Cleanliness, deleterious materials, particle shape and surface texture must be evaluated first, and some additional tests are run to determine gradation and size, specific gravity and absorption. At last, the gradations are selected and trial blends of these different gradations are calculated until an acceptable final mix design gradation is achieved.

(2) Asphalt Binder Evaluation

The grade of asphalt cement is selected based on the expected temperature range and traffic conditions.

(3) Sample Preparation

The full Marshall mix-design procedure requires 18 specimens 101.6 mm (4 in) in diameter and 63.5 mm (2.5 in) high. The stability and flow are measured for 15 specimens. In addition, three specimens are used to determine the theoretical maximum specific gravity. This value is needed for the void and density analysis. The specimens for the theoretical maximum specific gravity determination are prepared at the estimated design asphalt content. Based on the optimum asphalt binder content estimate, samples are typically prepared at 0.5 percent by weight of mix increments, with at least two samples above the estimated asphalt binder content and two below. Each sample is heated to the anticipated compaction temperature and compacted with a Marshall hammer.

The Asphalt Institute permits three different levels of energy to be used for the preparation of the specimens: 35, 50, and 75 blows on each side of the sample.

Most mix designs for heavy-duty pavements use 75 blows, since this better simulates the required density for pavement construction.

(4) The Marshall Stability and Flow Test

The Marshall stability and flow test provides the performance prediction measure for the Marshall mix design method. The stability is the maximum load supported by the test specimen. The test is performed at a deformation rate of 51 mm/min (2 in/min) and a temperature of 60 °C (140 °F). The Marshall flow is the deformation of the specimen when the load starts to decrease. Stability is reported in newton (pounds) and flow is reported in units of 0.25 mm (0.01 in) of deformation.

(5) Density and Voids Analysis

All mix design methods use density and voids to determine basic HMA physical characteristics. Two different measures of densities are typically taken:

① Bulk specific gravity (GMB): the ratio of the mass in air of a unit volume of a permeable material (including both permeable and impermeable voids normal to the material) at a stated temperature to the mass in air (of equal density) of an equal volume of gas-free distilled water at a stated temperature.

② Theoretical maximum specific gravity (TMD, GMM): the ratio of the mass of a given volume of void less ($V_a = 0$) HMA at a stated temperature (usually 25 °C) to a mass of an equal volume of gas-free distilled water at the same temperature.

These densities are then used to calculate the volumetric parameters of the HMA:

① Air voids (V_a), sometimes expressed as voids in the total mix (VTM):

$$V_a = \left(1 - \frac{G_{\text{mb}}}{G_{\text{mm}}}\right) \times 100\%$$

V_a is expressed as a percent of the bulk volume of the compacted paving mixture.

$$V_a = \frac{V_V}{V_T} \times 100\%$$

In this fornula: V_V is the volume of air voids;

V_T is the total volume of the mixture.

② Voids in the mineral aggregate (VMA).

③ Voids filled with asphalt (VFA).

Generally, these values must meet local or State criteria.

Table 7.6 Typical Marshall Minimum VMA

Nominal Maximum Size/mm		Minimum VMA/%
-	(U.S.)	
63	2.5 inch	11
50	2.0 inch	11.5
37.5	1.5 inch	12
25.0	1.0 inch	13
19.0	0.75 inch	14
12.5	0.5 inch	15
9.5	0.375 inch	16
4.75	No. 4 sieve	18
2.36	No. 8 sieve	21
1.18	No. 16 sieve	23.5

(6) Selection of Optimum Asphalt Binder Content

The optimum asphalt binder content is finally selected based on the combined results of Marshall stability and flow, density analysis and void analysis. Optimum asphalt binder content can be arrived at in the following procedure:

① Plot the following graphs:

■ Asphalt binder content vs. density. Density will generally increase with increasing asphalt content, reach a maximum, then decrease. Peak density usually occurs at a higher asphalt binder content than peak stability.

■ Asphalt binder content vs. Marshall stability. This should follow one of two trends: stability increases with increasing asphalt binder content, reaches a peak, then decreases; stability decreases with increasing asphalt binder content and does not show a peak. This curve is common for some recycled HMA mixtures.

■ Asphalt binder content vs. flow. Flow value is an index to evaluate the plastic deformation resistance of asphalt mixture.

■ Asphalt binder content vs. air voids. Percent air voids should decrease with increasing asphalt binder content.

■ Asphalt binder content vs. VMA. Percent VMA should decrease with increasing asphalt binder content, reach a minimum, then increase.

■ Asphalt binder content vs. VFA. Percent VFA increases with increasing asphalt binder content.

② Determine the asphalt binder content that corresponds to the specifications median air void content (typically this is 4 percent). This is the optimum asphalt binder content.

③ Determine properties at this optimum asphalt binder content by referring to the plots. Compare each of these values against specification values and if all are within specification, then the preceding optimum asphalt binder content is satisfactory. Otherwise, if any of these properties is outside the specification range the mixture should be redesigned.

2. Superpave Mix Design Method

Superpave is an acronym for Superior Performing Asphalt Pavements. The Superpave mix design method is one of the principal results from the Strategic Highway Research Program (SHRP). This method is a volumetric mix design process which accounts for traffic loading and environmental conditions. An analysis of specimens and the maximum specific gravity sample are conducted to evaluate such properties as voids in mineral aggregate (VMA); voids filled with asphalt (VFA), air voids, and the dust/effective binder ratio. The mix designer uses this information to determine the parameters that require adjustment before fabricating additional specimens. This process is repeated several times until the designed aggregate structure and the binder content produce specimens with the desired volumetric properties. Using the information obtained from this procedure, the mix designer then proceeds with preparing two specimens at four binder contents in preparation for determining the optimum binder content required to produce the four percent air voids at N_{des} gyrations.

The Superpave mix-design process consists of seven steps: (1) selection of aggregates, (2) selection of binder, (3) sample preparation (including compaction), (4) performance tests, (5) density and voids calculations, (6) determination of the design binder content, (7) evaluation of moisture susceptibility.

(1) Selection of Aggregates

The contractor who will be doing the paving sends WSDOT three trial aggregate blends (typically a coarse, fine, and middle ground gradation) along with laboratory data for each of these blends. The contractor indicates which of the gradations he/she would like to use and designates a design asphalt content. The aggregate requirements for Superpave are checked by the contractor during his/her trial blend process, then again by WSDOT during the confirmation of the contractor's proposal. The coarse aggregate angularity requirements are determined by the number of ESALs to which the roadway will be subjected. Additionally:

■ Unless the aggregate comes from a previously WSDOT-approved stockpile, testing is done to confirm the aggregates meet WSDOT specifications. As of 2002, aggregate sources are approved for 5 years, although some sources have not been switched over from the previous 10-year approval interval.

■ Each of these trial blends must be within the Superpave graduation requirements and preferably not pass through the restricted zone.

■ The bulk specific gravity (G_{sb}) of the coarse and fine aggregate is determined for each stockpile. In this case, material retained on the 4.75 mm (No. 4) is considered "coarse", while the rest is considered "fine".

(2) Binder Selection

The contractor who will be doing the paving sends WSDOT the brand and type of binder and antistrip modifier to be used. Actual asphalt binder samples are sent from the asphalt producers whenever needed throughout the year. Producers typically send anywhere from 10 to 40 cans at a time (depending on the binder type - typically they send fewer of the modified binders). The asphalt binder shall conform to AASHTO MP 1 requirements (Superpave PG binder system). WSDOT only allows one asphalt binder type submission for Superpave jobs. WSDOT determines the asphalt binder's specific gravity for use in the mix design process.

(3) Sample Preparation

Typically, six initial samples are made: two at the design asphalt content, two at 0.5 percent below the design asphalt content and two at 0.5 percent above the design asphalt content. These six samples are then cured and conditioned according to AASHTO PP 2 and compacted in the Superpave gyratory compactor in accordance with AASHTO TP 4.

Additionally, three samples (one at each of the above asphalt contents) are made and compacted in the California kneading compactor for use in stability tests.

- AASHTO PP 2: *Mixture Conditioning of Hot Mix Asphalt*
- AASHTO TP 4: *Method for Preparing and Determining the Density of Hot Mix Asphalt (HMA) Specimens by Means of the Superpave Gyratory Compactor*

(4) Performance Tests

The original intent of the Superpave mix design method was to subject the various trial mix designs to a battery of performance tests akin to what the Hveem method does with the stabilometer and cohesiometer, or the Marshall method does with the stability and flow test. Currently, these performance tests, which constitute the mixture analysis portion of Superpave, are still under development and review and have not yet been implemented. The most likely performance test, called the Simple Performance Test (SPT) is a Confined Dynamic Modulus Test.

(5) Density and Voids Analysis

First, bulk specific gravity (G_{mb}) is determined for each sample and the two results for each asphalt content are averaged. Second, one sample from each asphalt content is broken down for density and volumetric determinations to include theoretical maximum density (abbreviated TMD or called "Rice" density after its originator, and often designated G_{mm}), air voids, VMA and VFA. At this time the G_{mm} at $N_{initial}$, N_{design} and $N_{maximum}$ are checked, as well as the dust to asphalt ratio. The effective asphalt content (P_{be}) and percent absorbed asphalt content (P_{ba}) are also checked.

(6) Selection of Optimum Asphalt Binder Content

Using the data from the three asphalt contents, the optimum binder content is selected as that which corresponds to 4.0 percent air voids (4.5 percent air voids for Superpave designs that will be paid for based on volumetric properties). Usually, this asphalt content must be interpolated between two of the sample asphalt contents. For example, a 5.0 percent asphalt sample may have 4.8 percent air voids and a 5.5 percent asphalt sample may have 3.8 percent air voids. In this case the design asphalt content would be interpolated as 5.4 percent. This selected asphalt content must also meet VMA, VFA, density and dust-to asphalt requirements.

(7) Moisture Susceptibility Evaluation

Moisture susceptibility testing is the only performance testing incorporated in the Superpave mix design procedure as of early 2002. The modified Lottman test is used for this purpose.

7.7.4 Technical Properties of Asphalt Concrete

As a pavement material, asphalt concrete is bound to bear the repeated friction of the vehicles and the long-term impact of the environment. Therefore, an asphalt concrete should be designed, produced and placed in order to obtain the following desirable mix properties.

1. High Temperature Stability

Asphalt concrete is a kind of viscoelastic plastic material and its bearing capacity varies with temperature. The bearing capacity decreases as temperature rises. Especially under high-temperature or long-time bearing condition, the deformation occurs obviously. Part of the irrecoverable deformation becomes permanent deformation, which results in rutting, wave, upheaval and some other high temperature diseases. Rutting not only decreases the useful life of a pavement but also creates a safety hazard for the traveling public.

A standardized laboratory equipment and test procedure that predicts field-rutting potential would be of great benefit to the hot mix asphalt industry. Currently, the most common type of laboratory equipment of this nature is a loaded wheel tester (LWT). In China, the common LWT is developed from Japan. The external diameter and the width of the loaded tire is respectively 200 mm and 50 mm. The loading speed is 42 times/min and testing temperature is (60±1) °C. The specimen size is 300 mm×300 mm×50 mm. Dynamic stability (DS) is used as the evaluating index.

Several LWTs currently are being used in some developed countries, which include the Georgia Loaded Wheel Tester (GLWT), Asphalt Pavement Analyzer (APA), Hamburg Wheel Tracking Device (HWTD), LCPC (French) Wheel Tracker, Purdue University Laboratory Wheel Tracking Device (PUR Wheel), and one-third scale Model Mobile Load Simulator (MMLS3).

2. Low Temperature Performance

Low temperature cracking is also the main distress in asphalt pavements in winter. Under low temperature condition, the volume of the asphalt concrete will shrink, but it cannot shrink freely because of the restrain of the surrounding materials, which results in temperature stress from the interior of the materials. As the temperature drops sharply in winter, the temperature stress cannot relax timely. when the stress exceeds the allowable value of the asphalt mixture, the crack will occur, that is low temperature damage of the asphalt pavement is caused.

The evaluation methods of low temperature performance of asphalt mixture are not unified. Currently, the methods include three types: predicting the cracking temperature, evaluating low temperature deformation ability and evaluating rupturing ability of asphalt concrete. The most common used method in the world is the thermal stress restrained specimen test (TSRST) from Sharp research program. In China, beam bending test is usually performed to evaluate the low-temperature deformation ability of asphalt concrete.

3. Durability

The durability of an asphalt concrete is the ability to resist factors such as changes in the binder (polymerization and oxidation), disintegration of the aggregate, and stripping of the binder films from the aggregate. These factors can be the result of weather, traffic, or a combination of the two. Generally, durability of a mixture can be enhanced by three methods. They are: using maximum binder content, using a dense gradation of stripping-resistant aggregate, and designing and compacting the mixture for maximum impermeability.

4. Workability

Workability describes the ease with which a paving mixture can be placed and compacted. Mixtures with good workability are easy to place and compact; those with poor workability are difficult to place and compact. Workability can be improved by changing mix design parameters, aggregate source, and gradation.

5. Skid Resistance

Skid resistance is the ability of an asphalt surface to minimize skidding or slipping of vehicle tires, particularly when wet. It is related to the roughness, gradation, binder content and the surface properties of the aggregates. Using hard and angular rock, increasing particle size, reduce the binder content properly and some other measures can increase the skid resistance of the asphalt pavement.

6. Impermeability

Impermeability is the resistance of an asphalt pavement to the passage of air and water into or through it. This characteristic is related to the void content of the compacted mixture, and much of the discussion on voids in the mix design sections relates to impermeability. Even though void content is an indication of the potential for passage of air and water through a pavement, the character of these voids is more important than the number of voids. The size of voids, whether or not the voids are interconnected, and the access of the voids to the surface of the pavement all determine the degree of impermeability.

7.7.5 Reclaimed Asphalt Concrete

1. Hot Recycling

Reclaimed asphalt pavement (RAP) can be used as an aggregate in the hot recycling of asphalt paving mixtures in one of two ways. The most common method (conventional recycled hot mix) involves a process in which RAP is combined with virgin aggregate and new asphalt binder in a central mixing plant to produce new hot mix paving mixtures. A second method (hot in-place recycling) involves a process in which asphalt pavement surface distress is corrected by softening the existing surface with heat, mechanically removing the pavement surface, mixing it with a recycling or rejuvenating agent, possibly adding virgin asphalt and/or aggregate, and replacing it on the pavement without removing the recycled material from the pavement site.

Reclaimed asphalt pavement must be processed into a granular material prior to use in hot mix applications. A typical RAP processing plant consists of a crusher, screening units, conveyors, and stacker. It is desirable to produce either a coarse or a fine fraction of processed RAP to permit better control over input to the hot mix plant and better control of the mix design. In the hot in-place recycling (HIPR) process, the surface of the pavement must be softened with heat prior to mechanical scarification. The HIPR process has evolved into a self-contained, continuous train operation that includes heating, scarifying, rejuvenator addition, mixing, and replacement.

2. Cold Recycling

Reclaimed asphalt pavement (RAP) can be used as an aggregate in the cold recycling of asphalt paving mixtures in one of two ways. The first method (cold mix plant recycling) involves a process in which RAP is combined with new emulsified or foamed asphalt and a recycling or rejuvenating agent, possibly also with virgin aggregate, and mixed at a central plant or a mobile plant to produce cold mix base mixtures. The second, more common, method involves a process in which the asphalt pavement is recycled in-place [cold in-place recycling (CIPR) process], where the RAP is combined without heat and with new emulsified or foamed asphalt and/or a recycling or rejuvenating agent, possibly also with virgin aggregate, and mixed at the pavement site, at either partial depth or full depth, to produce a new cold mix end product. Most constructor have used cold in-place recycling in conjunction with a hot mix overlay or chip seal.

Processing requirements for cold mix recycling are similar to those for recycled hot mix. Recycled asphalt pavement must be processed into a granular material prior to use in cold mix applications. A typical RAP plant consists of a crusher, screening units, conveyors, and stackers. Cold in-place recycling(CIPR), like hot in-place recycling (HIPR), requires a self-contained, continuous train operation that includes ripping or scarifying, processing (screening and sizing/crushing unit), mixing of the milled RAP, and the addition of liquid rejuvenators. Special asphalt-derived products such as cationic, anionic, and polymer modified emulsions, rejuvenators and recycling agents have been developed especially for CIPR processes. These hydrocarbon materials are sometimes, but not always, used to soften or lower the viscosity of the residual asphalt binder in the RAP material so that it is compatible with the newly added binder.

7.7.6 Warm Mix Asphalt Concrete

Warm Mix Asphalt (WMA) is the generic term for a variety of technologies that allow producers of Hot Mix Asphalt (HMA) pavement material to lower temperatures at which the material is mixed and placed on the road. It is produced by adding either zeolites, waxes, asphalt emulsions, or sometimes even water to the asphalt binder prior to mixing. This allows significantly lower mixing and laying temperatures and results in lower consumption of fossil fuels, thus releasing less carbon dioxide, aerosols and vapors. Not only are working conditions improved, but the lower laying-temperature also leads to more rapid availability of the surface for use, which is important for construction sites with critical time schedules. The usage of these additives in hot mixed asphalt (above) may afford easier compaction and allow cold weather paving or longer hauls.

Questions

7.1 What is asphalt? What is tar?

7.2 Describe the four fractions of asphalt and their characteristics.

7.3 What is physical properties of asphalt? How to calculate penetration index?

7.4 Define cutback asphalt and emulsified asphalt. How to grade emulsified asphalt?

7.5 What are the differences between asphalt concretes with different composition of the aggregate gradation?

7.6 How to reduce the consumption of fossil fuels when producing WMA?

References

[1] Asphalt Institute. Asphalt cold-mix recycling[M]. Lexington, KY: Asphalt Institute, 1983.

[2] Asphalt Institute. Asphalt Hot-Mix Recycling, Manual Series No.20[M]. 2nd ed. Lexington, KY: The Asphalt Institute, 1986.

[3] ROBERTS F L, KANDHAL P S, BROWN E R, et al. Hot mix asphalt materials, mixture design and construction [J]. Asphalt Mixtures, 1991.

[4] BUTTON J W, LITTLE D N, ESTAKHRI C K. Hot in-place recycling of asphalt concrete [M]. Washington, DC: Transportation Research Board, 1994.

[5] JON A EPPS. Cold-Recycled Bituminous Concrete Using Bituminous Materials[M]. Washington, DC: Transportation Research Board, 1990.

[6] L A COOLEY JR, P S KANDHAL, M S BUCHANAN, et al. Loaded Wheel Testers in the United States: State of the Practice [M]. Washington, DC: Transportation Research Board, 2000.

[7] MICHAEL S MAMLOUK, JOHN P ZANIEWSKI. Materials for Civil and Construction Engineers [M]. Stockton: QWE Press, 2011.

[8] ROBERTS F L, KANDHAL P S, BROWN E R, et al. Hot mix asphalt materials, mixture design and construction (2rd)[J]. Construction Equipment, 1996.

Chapter 8
Wood

8.1 Introduction

Wood, because of its availability, relatively low cost, ease of use, and durability, continues to be an important civil engineering material. Wood is used extensively for buildings, bridges, utility poles, floors, roofs, trusses, and piles. Civil engineering applications include both natural wood and engineered wood products, such as laminates, plywood, and strand board.

Wood is used as architectural and decoration material for its several advantages as follows: high specific strength (lightweight and high strength), high frost resistance, chemical resistance, low thermal conductivity, easy processing, easy components connection. And for sure the wood also has following disadvantages: uneven structure, natural flaw, high moisture absorption, which lead to larger size change, easier warping, cracking, burning and erosion. But if we can use reasonably, the above disadvantages can be overcome.

Trees are a kind of natural resources with slow growth. However, there is a great demand for trees in various industries. In China, the forest coverage rate is about 16.55%, while the world average forest coverage rate is 31.7%. For civil engineering, it is important to understand the properties of wood correctly and make rational and economical use of wood resources.

8.1.1 Classification

According to tree species, trees are mainly classified into two species: conifer and broadleaf.

1. Conifer

The main characteristics of conifers are as follows: The leaves are slime and long and needle-like, the trunks are straight and tall, and ligneous tissue is soft, liable to process. Of superior strength, apparent density is low, and shrinkage deformation is low. The most of them used in architecture, mainly for doors and windows, decoration or bearing component.

The common species are pine, juniper, cypress, etc.

2. Broad-leaved Tree

The main characteristics of broad-leaved trees are as follows: Leaves are broad and shape in sheets, so most of which are hardwood. The straight parts of the trunks are short, and ligneous tissue is hard, not easy to process. The apparent density is high, and the shrinkage deformation is high, which is easy to crack or warp. This kind of trees are used for minor load-carrying member in interior decoration or veneer. The common species are birch, elm, etc.

8.1.2 Structure

The performance of wood derives from the structure of the wood. Civil and construction engineers need to understand the way the tree grows and the anisotropic nature of wood in order to properly design and construct wood structures. The structure of wood can be classified into macrostructure and microstructure.

1. The Macrostructure of Wood

The wood structures that can be seen by eyes or through magnifying glass are called the macro structure of wood.

In order to observe closely, the trunks are cut into three different sections:

- Transverse section: the section that is vertical against the trunk axis;
- Radial section: the section that passes the trunk axis;
- Tangential section: the section that parallels with the trunk axis and tangent with the annual ring.

The wood is made up of bark, xylem and pith. Bark is mainly used for burning except certain species of trees (cork oak, and yellow pineapple tree) whose bark can be used to make heat-proof materials. Pith is in the central part of the trunk, whose texture is loose and fragile, and is liable to be corrupted or eaten by insect worms. So the best part for use is the xylem of trunk.

In the xylem, the darker part near the pith is called duramen, while the light part outside is called albumum. The duramen contains little water so that it is not liable to reshape, and it has high corrosion resistance. While the alburnum contains more water that it is easy to deform and has worse corrosion resistance than duramen.

On the transverse section many centric circles can be seen, which are called annual rings. Of them the part in dark color and lie close are grown in summer, called summerwood. And the converse part is grown in spring, called springwood. The more summerwood wood has, the better the wood is. The more intense and evener annual rings the wood has, the better quality the wood has.

2. The Microstructure of Wood

The wood structures that can be seen through microscope are called microstructure of wood. Wood is composed of numerous tubular cells. The long dimension of the majority of cells is parallel to the tree's trunk. However, a few cells, in localized bundles, grow radially, from the center to the outside of the trunk.

Every cell can be classified into two parts: cell wall and the lumen. The cell wall is composed of fibrils. The longitudinal combination is firmer than the transverse combination. So the cell wall is of high strength in lengthways, but of low strength in transverse. There are very little spaces among the fibrils composed of cell wall, which enables the material to absorb or leak water.

The structure of the cell determines the physical characteristics of wood. For example: the wood with thick cell wall and small lumen is intense and hard, and its bulk specific gravity is high and it is of high strength.

8.2 Properties of Wood

8.2.1 Physical Properties

1. Moisture Content of Wood

The moisture of wood is measured in the percentage of water content, which is the percentage of the mass of water to the mass of dry wood.

(1) The Water in Wood

The water in the wood can be classified into the free water that lies in intercellular space and the absorbed water that lies inside the cell wall. The newly-cut wood is green wood. There is a plenty of free water and absorbed water in it. And the percentage of water content ranges from 70 % to 140%. When wood becomes dry, the free water is the first to evaporate, but at this time the size and mechanical property of wood are not influenced. When the free water finishes evaporating, the absorbed water begins to evaporate. The process of absorbed water evaporating is slow, and during it the bulk and the strength change regularly.

(2) Fiber Saturation Point

The status when there is no free water in wood, but the cell walls are saturated with absorbed water, is called the fiber saturation point. In general, the fiber saturation point of wood is from 25% to 35%.

(3) Equilibrium Water Content

The status that the percentage of water content of wood keeps balance with the surrounding moisture is called equilibrium water content. In order to avoid deformation and splits of wooden products caused by the change of moisture of wood, the wood must be dried until the percentage of water content reaches the equilibrium water content. In the north area of China, the equilibrium water content is about 12%, while in the south area the balanced percentage of water content is 15%-20%. The kiln-dried wood's percentage of water content is 4%-12%.

2. Dry Shrinking and Wet Swelling

When absorbed water content in cell walls changes, the deformation of wood may arise, which is wet swelling and dry shrinking. During the process that wood is dried from damp status to cellar saturation point, the size of wood remains still but mass decreases. Only when wood remains being dried until the absorbed water in cell wall begins to evaporate, do the wood begin to shrink. And when the absorbed water in wood begins to grow, the wood will start to expand.

Because the structure of wood is not even, the shrinking and swelling also varied from direction to direction. The shrinkage value is the smallest in the direction of long grain, and bigger in the radial direction, and the most in the chord direction. When wood becomes dry, the size and the shape of section may change a lot.

The shrinkage effect makes a great difference to the usage of wood. It may cause the wood split or warp, even make the structure of wood loosen or heave. The most fundamental measure to avoid these negative impact is to dry the wood before processing to keep the moisture content of the wood in balance with the humidity around the wood parts.

8.2.2 Strength and Affecting Factors

1. The strength of wood

Strength properties of wood vary to a large extent, depending on the orientation of grain relative to the direction of force. According to the ways that wood bears force, the strength of wood can be classified into tensile strength, compression strength, bending strength and sharing strength. And the tensile strength, compression strength and sharing strength also vary with the parallel grain (the direction of force parallels with the fiber direction) and transverse grain (the direction of force is vertical against the fiber direction). The parallel grain strength is quite different from the transverse grain strength. According to the Table 8.1, you can see how to make good use of all species of wood on the basis of their strengths separately.

Table 8.1 The relationships between strengths of wood

Compression strength		Tensile Strength		Bending strength	Shear Strength	
parallel grain	transverse grain	parallel grain	transverse grain		parallel grain	transverse grain
1	1/10-1/3	2-3	1/20-1/3	3/2-2	1/7-1/3	1/2-1

In Chinese construction projects, the physical and mechanical properties of the commonly used woods are shown in Table 8.2.

Table 8.2 The physical and mechanical properties of commonly used wood

Name of Wood	Place of Production	Air Dry Density /(kg/m^3)	Parallel Grain Compression Strength /MPa	Parallel Grain Tensile Strength /MPa	Bending Strength /MPa	Parallel Grain Shear Strength/MPa	
						Radial plane	Chord plane
Coniferous wood:							
Cedar wood	Hunan	371	38.8	77.2	63.8	4.2	4.9
	Sichuan	416	39.1	83.5	68.4	6.0	5.9
Red pine	Northeast	440	32.8	98.1	65.3	6.3	6.9
Masson pine	Anhui	533	41.9	99.0	80.7	7.3	7.1
Dahurian larch	Northeast	641	55.7	129.9	109.4	8.5	6.8
Picea-jezoensis	Northeast	451	42.4	100.9	75.1	6.2	6.8
Weeping cypress	Hubei	600	54.3	117.1	100.5	9.6	11.1
Broad-leaved wood							
Toothed oak	Northeast	766	55.6	155.1	124.1	11.8	12.9
German oak	Anhui	930	52.1	155.4	128.6	15.9	18.0
Fraxinus mandshurica	Northeast	686	52.5	138.1	118.6	11.3	10.5
Poplar	Shanxi	486	42.1	107.0	79.6	9.5	7.3

In accordance with *Wood physical performance mechanics test methods*, the spotless standard specimens are used to test wood strength. In experiment, there are different damaging conditions when wood suffers from different external forces. Parallel grain compression damage is caused by loss of stability of cell walls, not fiber breakage; Transverse grain compression damage is caused by significant deformation after compression; Parallel grain tensile damage is caused by tear among fibers and then tensile failure.

Due to the differences of force on wood fiber direction, the status when wood is sheared can be classified into parallel grain shear, transverse grain shear and transverse grain cutting. Parallel grain shear damage is caused by longitudinal displacement and transverse grain tension resulting from tear among the bonding of fibers; Transverse grain shear damage is caused by the tear among transverse bonding of fibers in shear plane; Transverse grain cutting damage is caused by fibers being cut and the strength at present is about 3 to 4 times than that of parallel grain shear.

2. Factors Affecting the Wood Strength

Besides its own structure, the strength of wood is also determined by such factors as the percentage of wood moisture, the defects (knots, irregular grain, splits, decay rot and worm rot), the duration of outside force and temperature.

(1) Water Content of Wood

Strength of the wood is greatly affected by water content. When the wood contains less water than the saturation point, the percentage of moisture reduces, and the absorbed water becomes less and less, so that the strength of wood rises. To the contrary, the absorbed water increases and the cell walls expand, then the structure loosens and the strength of wood lowers. When the percentage of moisture exceeds the fiber saturation point, only free water is changing, and the strength of wood remains still.

The national standard *Method of Testing in Compressive Strength Paralled to Grain of Wood* (GB 1935-2009) provides standard strength value when moisture content is 12%. The strength of other moisture content shall be converted by the following formula:

$$\sigma_{12} = \sigma_{W}[1 + \alpha(W - 12)]$$

In the formula: σ_{12} the strength of 12% moisture content (MPa);

σ_W the strength of W% moisture content (MPa);

W moisture content (%);

α the coefficient of moisture content, when the water content is 9%-15%, the numeral values are determined according to Table 8.3.

Table 8.3 Moisture correction factor

Strength Type	Compression Strength		Tensile Strength Parallel to Grain		Bending Strength	Sharing Tensile Strength Parallel to Grain
α	Parallel grain	Transverse grain	Broadleaf wood	Conifer wood		
	0.05	0.045	0.015	0	0.04	0.03

(2) Environment Temperature

Temperature has direct influence on the wood strength. The experiment shows that when the temperature rises from 25 °C to 50 °C , the wood compression strength will be reduced by 20%-40% and the wood sharing strength will be reduced by 12%-20% because the collide among wood fibers is softened. In addition, if the wood is in hot and dry surrounding, it may become fragile. During the processing of wood, boiling method is often employed to reduce its strength contemporarily to meet the needs of processing (such as the production of plywood).

(3) The Duration of Outer Force

The limit strength of wood stands for the capability of standing the outer force in a short time. The limit that the wood can stand in a long run is the rupture strength of wood. Because plastic-flow deformation will occur to wood, the strength of wood will be reduced with the lasting of loading time, and the rupture strength of wood may be only 50%-60% of the limit strength of wood.

(4) Defects

The wood strength is judged by the samples without defects. In fact, during the growing, cutting and processing process of wood, there may be such defects as knots, splits and worm rot. These defects make the wood uneven, and destroy wood structures; all these influences may reduce the strength of wood, especially the tensile strength and the bending strength.

Besides the factors above, the species of trees, growing surroundings, the age of trees, and different parts of trees all influence the wood strength.

8.3 Applications of Wood in Architecture

During the construction process, the wood should be used rationally according to the species, the grade and the structure. And we should also try to avoid using the big ones for fraction and the good ones for trifles.

8.3.1 Species and Specifications of Wood

The wood used in architecture can be classified into primitive streak, log, sawn timber and crosstie according to its usage and status of processing.

Primitive streak means the wood without bark, root and treetop. And usually it is not processed into certain length or diameter by certain size. Primitive streak is often used as scaffold, architecture material and furniture.

Log means the wood without bark, root and treetop. And usually it is processed into certain length or diameter according to certain size. Log is often used as frame, purlin or rafter, etc. Furthermore, it can also be used as pile timber, pole, mine timber, etc. When processed, it can also be made into plywood, ship model and machine model.

Sawn timber means timber, which has been processed and sawn. The timber whose width is three (or more than three) times of its thickness is called plate. While the timber whose width is less than three times of its thickness is called square log. Sawn timber is often used in architecture, bridge, furniture, ship, automobile or pocking box.

Crosstie means the timber processed according to the section and length of sleeper. Crosstie is often used in railway construction.

8.3.2 Engineering Applications

Wood has the following advantages: high specific strength (light weight and high strength), high frost resistance, chemical resistance, low thermal conductivity, easy processing, easy components connection. As wood is flammable, special attention should be paid to wood processing. The properties of wood decide the utility scope of the wood.

In structure, wood is mainly used as frame and roof, but it is less used in modern buildings. Many ancient buildings have a high level of technology and art unique styles.

Because woods have beautiful patterns and special gloss, after surface processing, wood is used in building interiors and decorations commonly.

In order to improve the utilization rate of wood, the corner and debris of the wood is fully used to make into plywood, fiberboard, particleboard, wood fiber board and other man-made sheet.

8.4 Corrosion and Fire Protection

8.4.1 Wood Decay and Anti-corrosion

The biggest drawback of wood is easy to decay, which leads to reducing the durability of wooden pieces and its products.

1. Wood Decay

The most serious destroyer of wood is rot fungi. It can decompose such as cellulose in the cell wall into simple materials as a fuel for their own breeding, which results in lumber decay. The conditions fit for rot fungus's survival and reproduction in wood: the suitable moisture, air and temperature. So keeping the wood dry frequently, it will not decay. What's more, that wood's fully immersed in water or buried deeply underground will also not decay caused by lack of air.

Besides the infringement of fungi, the wood could be corroded by insects such as termites and longicorns.

2. Wood Anticorrosion Measures

Two kinds of measures are adopted usually. One is to make the condition not suitable for parasitism and reproduction of fungus. The other one is to kill fungus or prevent fungal growth by drug.

Keeping the wood dry and making its moisture content less than 20%; Painting on the surface of wood and wood products in order to separate the air and water.

Using chemical preservatives for wood is a more effective corrosion protection measure.

There are many kinds of wood preservatives which generally divided into waterborne preservative, oily preservative and paste preservative. The preservative treatment methods of wood are surface coating or spray method, impregnation method, osmosis pressure method and hot and cold tank immersion method, etc.

8.4.2 Fire Protection of Wood

Flammability is the biggest disadvantage of wood, and the following methods are often adopted to protect wood from fire.

(1) Put the wood into flameproof infusion and ensure a proper amount of infusion and the filtration depth to meet the demand of fireproofing.

(2) Paint or spray the flameproof coating on to the surface of wood until the coating becomes dry. The flameproof effect depends on the thickness of coating or the amount used in every square meter.

With the passing of time and the influence of surrounding factors, the fireproof components in fireproof paints or infusion may decrease or decay that the fireproof ability of wood will decrease if the above two methods are adopted.

Questions

8.1 What are the two main classes of wood? What is the main use of each class? State the names of two tree species of each class.

8.2 Discuss the anisotropic nature of wood. How does this phenomenon affect the performance of wood?

8.3 What is equilibrium water content? What's the influence of dry shrinking and wet swelling?

8.4 Briefly describe water content of wood and its calculate method.

8.5 What kinds of wood are commonly used in engineer?

8.6 Introduce the wood anticorrosion measures and their principles.

References

[1] HAIMEI ZHANG. Building Materials in Civil Engineering[M]. Beijing: Science Press, 2011.

[2] MICHAEL S MAMLOUK, JOHN P ZANIEWSKI. Materials for Civil and Construction Engineers [M]. Stockton: QWE Press, 2011.

Chapter 9
Synthetic Polymers

9.1 Introduction

Polymers are long continuous chain molecules formed from repeated sequences of small organic units, whose molecular weight is in excess of 10000. Polymers include natural polymers and synthetic polymers. Natural polymers occur in nature and can be extracted. They are often water-based. Synthetic polymers, most of which were developed in just the last 60 or so years, include plastics, synthetic rubbers, synthetic fibers and adhesive. Synthetic polymers have been widely used in civil engineering for many years, such as material modification, construction decoration, waterproof, adhesive, anticorrosion, etc.

1. Plastics

Plastics are classified into two categories according to what happens to them when they're heated to high temperatures: thermoplastics and thermosets.

Thermosetting plastics are polymer materials that cannot be reformed after manufacturing cross linked chain networks. Because of their tightly crosslinking structure, they can resist higher temperatures and provide greater dimensional stability than most thermoplastics. Thermosets are tough, less creep, durable with high temperature performance, and have found applications in a wide variety of civil engineering including tubes, fiber-reinforced composites, polymeric coatings, adhesive, permanent structures, etc.

Thermoplastics are polymers that can be remolded after manufacturing. They soften upon reheating and have substantial creep, isotropic properties. These qualities also make thermoplastics recyclable. The civil engineering market is the second largest outlet for thermoplastics. There are many application samples including pipes, wall materials, coating, adhesive, modification agent for concrete and so on.

2. Synthetic Rubbers

Synthetic rubber is synthesized from petroleum and is classified as an artificial elastomer. This means that it is able to be deformed without sustaining damage, and can return to its original shape after being stretched. Manmade rubber has many advantages over natural rubber, and is used in many applications due to its superior performance. The use of synthetic rubber is much more prominent than natural rubber in most industrialized nations.

There are several different popular varieties of synthetic rubber. These are usually created by combining chemicals in different quantities during the rubber production process. The most prevalent synthetic rubbers are styrene-butadiene rubbers (SBR) derived from the copolymerization of styrene and 1,3-butadiene. Other synthetic rubbers are prepared from isoprene (2-methyl-1,3-butadiene), chloroprene (2-chloro-1,3-butadiene), and isobutylene (methylpropene) with a small percentage of isoprene for cross-linking.

Synthetic rubbers are widely used in civil engineering, such as water proof rolls, tube, asphalt modifier, additive for concrete.

3. Synthetic Fibers

Synthetic fibers are made from synthesized polymers or small molecules. They are manufactured using plant materials and minerals: viscose comes from pine trees or petrochemicals, while acrylic, nylon and polyester come from oil and coal.

Fibers and fabrics play a large role in everyday applications. A fiber is a hair-like strand of material. They are the smallest visible unit of a fabric and are denoted by being extremely long in relation to their width (at least 100 times longer than it is wide). Fibers can be spun into yarn and made into fabrics.

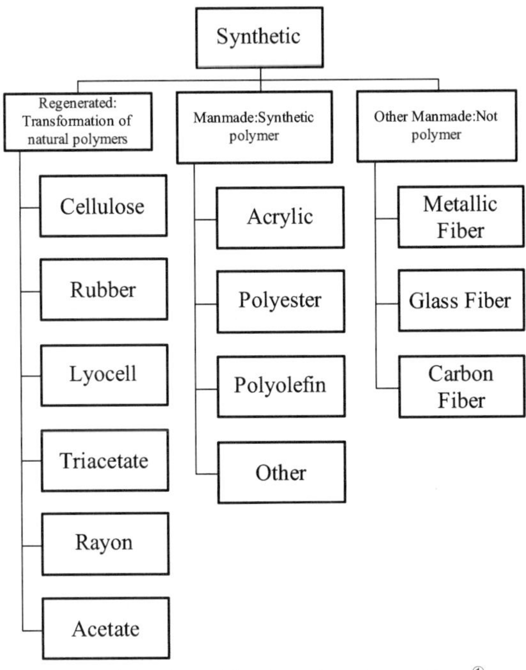

Figure 9.1 Classification of synthetic fibers[①]

① Engineering 360. Synthetic Fibers and Fabrics Information [OL]. [2020-10-01] http://www.globalspec.com/learnmore/materials_chemicals_adhesives/composites_textiles_reinforcements/synthetic_fibers_fabrics_polymer_textiles.

A great advantage of synthetic fibers is that they are more durable than most natural fibers. In addition, many synthetic fibers offer consumer-friendly functions such as stretching, waterproofing and stain resistance.

Synthetic fibers and fabric play a large role in every application. In civil engineering, they are mainly used in fiber reinforced concrete, structural reinforcement, geotextile and modifier for asphalt mixture.

4. Adhesive

Adhesive, any substance that is capable of holding materials together in a functional manner by surface attachment that resists separation. Adhesives mainly include three types: inorganic adhesive like cement, natural adhesive such as animal glue, starch, dextrin and natural gums, synthetic adhesive, such as epoxy, polyurethane, cyanoacrylate and acrylic polymers.

Although natural adhesives are less expensive to produce, most important adhesives are synthetic. Adhesives based on synthetic resins and rubbers excel in versatility and performance. Synthetics can be produced in a constant supply and at constantly uniform properties. In addition, they can be modified in many ways and are often combined to obtain the best characteristics for a particular application.

There are some examples of civil engineering application: attachment of materials to floor and walls, wooden and flexible floor covering, thermal insulation materials, components connection, reinforcement and repairs, and bonding layer for pavement.

An important challenge facing adhesive manufacturers and users is the replacement of adhesive systems based on organic solvents with systems based on water. This trend has been driven by restrictions on the use of volatile organic compounds (VOC), which include solvents that are released into the atmosphere and contribute to the depletion of ozone. In response to environmental regulation, adhesives based on aqueous emulsions and dispersions are being developed, and solvent-based adhesives are being phased out.

5. Coating

A coating is a covering that is applied to the surface of an object, usually referred to as the substrate. The purpose of applying the coating may be decorative, functional, or both. In general, coating consists of polymeric binder blended with functional pigments and other additives.

According to the solvent type, coating is classified into two categories: solvent-based coating and water-based coating. Solvent paints use a resin to hold the pigment in suspension and to help the pigment stick to the substrate. These solvents can release Volatile Organic Compounds (VOCs) into the atmosphere, with a negative and toxic impact on both environment and human beings. On the other hand, water-based coatings use few or no solvents but rather water. Generally, water dilutes the pigment and carries the pigment to the substrate. Therefore, water-based coatings could reduce the VOC emissions and are more environmentally friendly compared to solvent-based coatings.

Coating can change the surface properties of the substrate, such as adhesion, wettability, reflectivity, corrosion resistance, wear resistance, magnetic response and electrical conductivity, etc. Coating is widely applied in civil engineering, such as waterproofing coating for roof or ground, fireproof coating for steel structure, timber structure and concrete, wall coating, anticorrosive coating for steel structure, and reflective coating for building and pavement.

9.2 Common Polymers

9.2.1 Epoxy resin

Epoxy resin is mainly used as adhesive in civil engineering applications. It has excellent bonding properties and is suitable for the construction, repair, and rehabilitation of our transportation infrastructures.

Epoxy system are highly valued in these projects because they can be formulated to set outdoors in a relatively short period of time, they have good adhesion to a number of materials of construction, and they provide relatively good weathering properties and durability.

Table 9.1 lists several applications for epoxy adhesives in civil engineering projects. Epoxy adhesives are generally considered to be the "workhorse" of the industry. In the construction or repair of roads and bridges, epoxy adhesives have primarily been used for bonding concrete and for bonding stiffening members or repair structures to degrading concrete facilities.

Table 9.1 Common applications and polymeric resins used in the construction and repair of the transportation infrastructure

No.	Common applications and polymeric resins used in the construction and repair of the transportation infrastructure	No.	Common applications and polymeric resins used in the construction and repair of the transportation infrastructure
1	Airfield runways and aprons	6	Underground tunnel construction joints and sections
2	Highway and bridge joints	7	Attaching metal studs in concrete
3	Bridge abutments	8	Bonding pancake lighting systems on airport runways, parking lots, etc.
4	Concrete lined canals, traverse and longitudinal joints	9	Bonding traffic markers on roadways and road dividers
5	Multilevel parking lot joints		

Epoxies are high priced, but they have better chemical resistance and durability than other adhesives, and they have dominated the market in outdoor applications. Significant advantages of the epoxy based adhesives are that they have no solvents and, therefore, exhibit little shrinkage. They cure relatively fast and are not as exposed to inclement weather as are slower curing systems. Their physical properties do not change significantly on aging in the field.

One problem with early epoxy formulations is that they cured to a relatively brittle material. By using reactive flexibility agents, such as polysulfide, epoxy adhesive formulators have obtained the flexibility required for many applications in this industry. Polyamides and even coal tars have also been used to provide flexibility to epoxy base resins.

In recent years, epoxy resins are used as new modifiers for mortar or concrete, such as epoxy mortar, epoxy resin concrete and epoxy asphalt concrete. Epoxy mortar is a repair mortar that combines a traditional Portland cement and sand mix with a two-component epoxy to create a strong, fast-curing concrete material. Epoxy resin concrete composed of epoxy resin, hardener, additives, fillers and aggregates is applied in concrete pavement repair and structural reinforcement. Epoxy asphalt concrete is a polymer concrete made from a slow curing two-phase binder that is mixed with asphalt concrete aggregates. It does not become brittle at low temperature and does not melt at high temperature. Its most common applications are: (1) as a thin overlay (3/4 to 2 inch thickness) lightweight wear course for repair, replacement or new construction and (2) as a paving surface for orthotropic steel decks where toughness and elastic composite behavior are critical. Epoxy Asphalt pavements have much higher stability, much less susceptibility to cracking, are less permeable and maintain skid resistance much longer than conventional asphalt pavements.[①]

9.2.2 Polyester

Polyester is a category of polymers which contains the ester functional group in their main chain. Depending on the chemical structure, polyester can be a thermoplastic or thermoset. thermoplastics such as polybutylene terephthalate (PBT) and polyethylene terephthalate (PET), and thermosets. Thermoplastic polyesters are one of the fastest growing of the engineering thermoplastics.

Polyester can also be classified as saturated and unsaturated polyesters. Saturated polyesters refer to that family of polyesters in which the polyester backbones are saturated. They are thus not as reactive as unsaturated polyesters. They consist of low molecular weight liquids used as plasticizers and as reactants in forming urethane polymers, and linear, high molecular weight thermoplastics such as polyethylene terephthalate. Usual reactants for the

① Home Steady. Advantages of Epoxy Paint [OL]. [2017-07-21] http://homesteady.com/12325813/advantages-of-epoxy-paint.

saturated polyesters are a glycol and an acid or anhydride.

Unsaturated polyesters refer to that family of polyesters in which the backbone consists of alkyl thermosetting resins characterized by vinyl unsaturation. They are mostly used in reinforced plastics. These are the most widely used and economical family of resins.

Polyesters have many advantages, such as good chemical resistance, high strength, good ductility, convenience of use and low price. The disadvantages and limitations of polyesters include marginal bond quality, more expansion and shrinkage than concrete.

When it comes to civil engineering, there are many application examples including floor coatings, adhesive bonder or sealer, binder for fiberglass or artificial wood, binder for polymer mortar, polyester fiber reinforced concrete, and fiberglass-polyester paving mat.

9.2.3 Acrylic resins

Acrylic resins are thermoplastics or thermosetting plastics, derived from acrylic acids. They are used in various applications from paints and coatings, glass sheets, tail lights, signboards, to adhesives. It is considered as a good substitute for glass due to its hardness. Poly methyl methacrylate (PMMA) is one of the most widely used polymers of acrylics.

One of the main characteristic features of PMMA is its high transparency. With its high weather resistance, it has been known to last over 30 years, and it does not easily turn yellow or crumble when exposed to sunlight.

9.2.4 SBS

Styrene-butadiene-styrene (SBS) block copolymer is a type of styrene block copolymer (SBC) composed of styrene and butadiene linked photopolymer blocks. SBS copolymers belong to the class of thermoplastic elastomers that possess the mechanical properties of rubber, and processing capabilities of thermoplastic. SBS offers excellent surface friction coefficient, little permanent deformation, great tensile strength, excellent low-temperature behavior, great workability and good electric property.

The compound is well suited for being used as a sealing material and an adhesive in hot melt processing. The polymer is extensively used in civil engineering applications such as asphalt modification, polymer modification, liquid seal materials, waterproof coatings and adhesives.

9.2.5 SBR

Styrene-butadiene rubber (SBR) is one of the most versatile copolymer rubber compounds. It consists of the organic compound styrene and the chemical butadiene, with the amount of butadiene usually being about three times more than the amount of styrene. SBR is a stable synthetic that is resistant to abrasion.

The elastomer is used widely in pneumatic tires, shoe heels and soles, gaskets and even chewing gum. It is a commodity material which competes with natural rubber. Latex (emulsion) SBR is extensively used in coated papers, being one of the cheapest resins to bind pigmented coatings. SBR is widely used in civil engineering applications, as a sealing and binding agent behind renders as an alternative to PVA, but is more expensive. In the latter application, it offers better durability, reduced shrinkage and increased flexibility, as well as being resistant to emulsification in damp conditions. SBR can be used in "tank" damp rooms or surfaces, a process in which the rubber is painted onto the entire surface (sometimes both the walls, floor and ceiling) forming a continuous, seamless damp-proof liner; a typical example would be a basement. SBR added to asphalt can greatly improve asphalt's performance in stability, permanence, viscosity, and resistance to aging.

9.2.5 Polyvinyl chloride, PVC

Polyvinyl chloride is produced by polymerization of the monomer vinyl chloride (VCM). PVC is a thermoplastic polymer. Its properties are usually categorized based on rigid and flexible PVCs.

Its properties are usually categorized based on rigid and flexible PVCs.

Table 9.2 Properties of PVC

Property	Rigid PVC	Flexible PVC
Density/(g/cm^3)	1.3-1.45	1.1-1.35
Thermal conductivity/[W/(m · K)]	0.14-0.28	0.14-0.17
Yield strength/psi	4500-8700	1450-3600
Young's modulus/psi	490000	
Flexural strength (yield)/psi	10500	
Compression strength/psi	9500	
Coefficient of thermal expansion (linear) /[mm/(mm · °C)]	5×10^{-5}	
Vicat B/°C	65-100	Notrecommended
Resistivity/(Ω · m)	1016	1012-1015
Surface resistivity/Ω	1013-1014	1011-1012

PVC has high hardness and mechanical properties. The mechanical properties enhance with the molecular weight increasing but decrease with the temperature increasing. The mechanical properties of rigid PVC (uPVC) are very good; the elastic modulus can reach 1 500-3000 MPa. The soft PVC (flexible PVC) elastic is 1.5-15 MPa.

The advantages of PVC include excellent insulator, diverse applications, chemical resistance, long-term stability, flame resistant, weather resistant, adhesion to glass and resistance to oil, but it has low modulus and moisture sensitivity in production.

The excellent performance of PVC has resulted in it being used for a wide variety of civil engineering, such as pipe, window frames and moldings, floor tiles siding.

9.2.6 Polyethylene (thermoplastic)

Polyethylene (PE) is the most common plastic. The low cost of polyethylene production has encouraged producers to prefer its use over many other plastics. Its primary use is in packaging (plastic bag, plastic films, containers including bottles, etc.). Many kinds of polyethylene are known, with most having the chemical formula $(C_2H_4)_nH_2$.

Polyethylene is classified into several different categories based mostly on its density and branching. Its mechanical properties depend significantly on variables such as the extent and type of branching, the crystal structure and the molecular weight. With regard to sold volumes, the most important polyethylene grades are HDPE, LLDPE and LDPE.

In the last three decades, Polyethylene (PE) has become a widely used material in many of the civil engineering products, including various types of geosynthetics, pressure and corrugated pipe, and retention tanks. While products made from PE are lightweight, easy to install, and resist many chemicals, their mechanical and chemical properties can be significantly enhanced by adding a small amount of nanoclay, making PE/clay nanocomposites an attractive new class of engineering materials.[1]

9.2.7 Polyurethane

Polyurethane (PUR and PU) is a polymer composed of a chain of organic units joined by urethane links. Polyurethane polymers are traditionally and most commonly formed by reacting a di- or polyisocyanate with a polyol. Both the isocyanates and polyols used to make polyurethanes contain on average two or more functional groups per molecule.

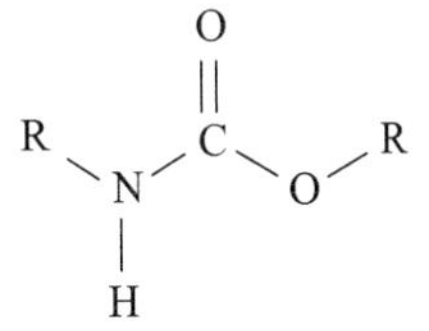

a urethane

Figure 9.2 The PU molecule

① Cairncross. Polyethylene Nanocomposites [OL]. (2012-05-02) [2010-09-09] http://www.chemeng.drexel.edu/cairncrossgroup/research/PN/default.aspx.

While most polyurethanes are thermosetting polymers that do not melt when heated, thermoplastic polyurethanes are also available. Polyurethanes are the most well-known polymers used to make foams. They are the single most versatile family of polymers there is. Polyurethanes can be elastomers, paints, fibers, and adhesives.

9.2.8 FRP

Polymer composites are multi-phase materials. By combining polymer resins such as polyester, vinyl ester and epoxy with fillers and reinforcing fibers to produce a bulk material, its properties are superior to that of the individual base materials. Fillers are often used to provide bulk to the material, reduce cost, lower bulk density or to produce aesthetic features. Fibers are used to reinforce the polymer and improve mechanical properties such as stiffness and strength. High strength fibers of glass, aramid and carbon are used as the primary means of carrying load, while the polymer resin protects the fibers and binds them into a cohesive structural unit. These are commonly called fiber composite materials.

Fiber-reinforced plastic (FRP) (also fiber-reinforced polymer) is a fiber composite material made of a polymer matrix reinforced with fibers. The fibers are usually glass, carbon, basalt or aramid, although other fibers such as paper or wood or asbestos have been sometimes used. The polymer is usually an epoxy, vinyl ester or polyester thermosetting plastic, and phenol formaldehyde resins are still in use.

Although the use of structural fiber composites in critical load-bearing applications is relatively rare one of its most common uses in the construction industry is repair of existing structures. The material is also used as a replacement for steel in reinforced and stressed concrete and in very rare cases to produce new civil structures almost entirely out of fiber composites.

Questions

9.1 How many kinds of plastics can be divided? What are their advantages?

9.2 What is the difference between synthetic fibers and natural fibers?

9.3 What is substrate? What are the advantages of water-based coating compared with solvent-based coating?

9.4 Briefly list the applications of common polymers.

9.5 Define fiber composite materials and FRP.

References

[1] G C MAYS， A R HUTCHINSON. Adhesives in Civil Engineering [M]. New York: Cambridge University Press, 1993.

[2] DEOPURA B, ALAGIRUSAMY R, JOSHI M, et al. Polyesters and Polyamides[M]. Abington: Woodhead Publishing, 2008.